Mathematics for the Majority

Some Routes through the Guides

Other titles in the series

MATHEMATICAL EXPERIENCE

MACHINES, MECHANISMS AND MATHEMATICS

ASSIGNMENT SYSTEMS

LUCK AND JUDGEMENT

MATHEMATICAL PATTERN

NUMBER APPRECIATION

MATHEMATICS FROM OUTDOORS

FROM COUNTING TO CALCULATING

ALGEBRA OF A SORT

GEOMETRY FOR ENJOYMENT

SOME SIMPLE FUNCTIONS

SPACE TRAVEL AND MATHEMATICS (2 volumes)

CROSSING SUBJECT BOUNDARIES

Mathematics for the Majority [1.]

Some Routes through the Guides

Chatto and Windus Educational
for
The Schools Council 1974

Granada Publishing Limited
First published in 1974 by Chatto and Windus Educational
Frogmore   St Albans   Hertfordshire   AL2 2NF
and
3 Upper James Street   London   W1R 4BP

ISBN 0 7010 0616 1

© Schools Council Publications 1974

All rights reserved. No part of this publication may be reproduced, stored in a retrieval system, or transmitted, in any form or by any means, electronic, mechanical, photocopying, recording or otherwise, without prior permission of the publishers.

Text set in 10/11 pt. IBM Univers, printed by photolithography, and bound in Great Britain at The Pitman Press, Bath

Contents

Part One

	Page
Chapter 1 Introduction	1
2 The writings — a short concordance	3
3 The mathematics curriculum	24
4 Some mathematics courses	30

Part Two

5 A book list	42
6 Some commercially produced materials with their sources	52

Part Three

7 Pupils at work — some topics for discussion by teachers	59

Acknowledgements

The tables on pp. 20 — 23 were reproduced by permission from the *Schools Council Examination Bulletin 25. CSE Mode 1 Examinations in Mathematics*, published for the Schools Council by Evans/Methuen Educational, 1972.

Mathematics for the Majority

The Schools Council Project in Secondary School Mathematics (now called *Mathematics for the Majority*) was set up to help teachers construct courses for pupils of average and below-average ability, to relate mathematics to their experience, and to provide them with some insight into the processes that lie behind the use of mathematics as the language of science and as a source of interest in everyday things.

Members of the Project Team (1969—1972)
- P. J. Floyd (Director)
- J. H. D. Parker (Deputy Director)
- K. C. Bonnaud
- T. M. Murray-Rust
- E. T. Norris
- Mrs. J. Stephens
- M. J. Cannon (Evaluator 1971/72)
- P. A. Kaner (Evaluator to 1971)

The chief author of this book is:
- P. J. Floyd

1 Introduction

> *The whole of mathematics consists in the organisation of a series of aids to the imagination in the process of reasoning'*
> A. N. Whitehead, *Universal Algebra* (Cambridge 1898).

It is of some significance that this book opens with the quotation which formed part of the finale of *Working Paper No 14 Mathematics for the Majority* (HMSO 1967), for it is on the contents of that Working Paper that the series of Guides which now comes under discussion was subsequently based.

The title, 'Some Routes through the Guides', has been chosen to convey the flavour of what this book attempts to do. It will list the series of Teachers' Guides published under the auspices of the Schools Council Mathematics for the Majority Project, and it will indicate briefly the contents of each Guide. Some policies for constructing a balanced and a relevant mathematics course for secondary pupils of average and below-average ability in mathematics will come under consideration, as will also some criteria for use in choosing a curriculum in mathematics. In attempting this, it will be inevitable that some sequences illustrating various facets will appear. These will exemplify some ways in which the Guides can be used in constructing courses. To provide definitive all-embracing courses is not our prime objective. We intend to give some positive leads in what we believe to be the right direction. If what appears here proves to be of direct use in some circumstances, then so much the better, that would constitute a bonus.

The responsibility for choosing a curriculum suited to a particular area must lie fairly and squarely on the shoulders of the teachers and other educators in that area. Any *ex cathedra* statement purporting to have universal application usurps the authority and responsibilities of those in the field, and as such, might fairly be dismissed as a professional impertinence. Our examples then will be chosen to help teachers in the problem of selection; this is a Guide among Guides, and as such it is intended, together with the others in the series, to be of help and guidance to those who have to make decisions in these matters — the teachers in a given locality.

Since so much continuing curriculum development at grass-roots level is conducted in Teachers' Centres and like places, some topics for discussion in these places will be given.

Part Two of this book will consist of a short (non-exhaustive) source-guide of books and materials which some teachers have found to be of value in their work.

From time to time, brief references will be made to the work of the Mathematics for the Majority Continuation Project. At the time of writing, that Project is in the early stages of producing packs of material for pupil use. This material is very broadly based on the work of the parent project, and it is intended for the use of average and below average ability pupils in the 13—16 year age band. The products of the Continuation Project will clearly have an important bearing whenever resources for implementing the work of the guides come under consideration.

As a Guide among Guides, the greater part of this book cannot be taken in isolation; it must be read in conjunction with whatever set of Guides and materials is considered relevant to the situation being considered.

This book was written after the draft versions of the Guides had been produced, but before the publication of the complete set. References to particular guides will therefore be given by *chapter only* rather than by page references in the published editions.

In the next chapter we consider the content of the Guides which have been produced by the Mathematics for the Majority Project.

2 The writings—a short concordance

'The person who has nowhere to go can never claim to have lost his way'
Space Travel and Mathematics Volume 2: Further Developments

The region through which we seek a number of pleasant paths consists of 14 published books, this book constituting the 15th of the series. Each book is a separate entity, so that, with one exception, no sequential development from one to another will be found in the set. The exception is the case of *Space Travel and Mathematics* which appears as two volumes, the second being a sequel to the first.

Without drawing boundaries which are too hard and fast, the books may be sorted into four sets.

1. Books which deal with fundamental mathematical ideas.
    From Counting to Calculating [CC]
    Number Appreciation [NA]
    Algebra of a Sort [AL]
    Some Simple Functions [SF]
    Geometry for Enjoyment [GE]

2. Books which present mathematics in action.
    Machines Mechanisms and Mathematics [MM]
    Luck and Judgement [LJ]
    Mathematics from Outdoors [MO]
    Space Travel and Mathematics Volume 1 [ST(1)]
    Space Travel and Mathematics Volume 2 [ST(2)]

3. Books which emphasize the pervasiveness and universality of mathematics.
    Mathematical Pattern [MP]
    Crossing Subject Boundaries [CB]

4. Books which highlight ways and means of organizing and presenting the subject.
    Mathematical Experience [ME]
    Assignment Systems [AS]
    Some Routes through the Guides [RG]

For the purposes of concise referencing later in this book, each of the titles is listed with a code reference. This code will be used to denote a particular guide, and the chapter in that guide to which attention is being drawn. For example:

(LJ 4) refers to the fourth chapter of *Luck and Judgement*.
(ST(2) 7) refers to the seventh chapter of *Space Travel and Mathematics Volume 2*.

Such a broad and rather arbitrary classification does little more than provide general directions, rough compass bearings to guide us in our travels. A closer analysis would reveal a considerable degree of overlap in many instances. In other words, the sets intersect. For example *Crossing Subject Boundaries* might equally well be regarded as mathematics in action. Whilst *Machines Mechanisms and Mathematics* provides a wealth of mathematics in action, it also devotes considerable space to the presentation of some of its ideas. This can also be said for the guides *From Counting to Calculating, Number Appreciation, Luck and Judgement, Mathematics from Outdoors,* and both books of *Space Travel and Mathematics*. This then is the justification for our earlier refusal to draw hard and fast lines when making our broad classification.

The team of the Continuation Project has produced packages of material for the use of pupils. These packages relate, in the main, to some environmental topics. They are published by Schofield and Sims.

There will thus be these resources to help teachers to implement the plans they make for the fuller mathematical education of older secondary pupils of average and below average ability.

Now for a more detailed but nevertheless brief consideration of the contents of the 14 books listed earlier, remembering that there is no pecking order in the list; the order in which they are taken will follow the list given earlier.

From Counting to Calculating [CC]. A study in arithmetic for secondary pupils

This book seeks to achieve four main objectives. They are:
(a) to help teachers to build up criteria appropriate to their own pupils, schools and localities, on which to base a programme of work on calculation;
(b) to suggest some fields of application appropriate to the needs and abilities of less able older secondary pupils;
(c) to pinpoint some of the ideas underlying computational methods in order to help the recognition of individual difficulties;
(d) to survey the use and introduction of suitable calculating aids and of devices which reduce the need for calculating.

The contents of the book are therefore in part philosophical and in part 'practical'.

The early chapters discuss, sometimes provocatively but always sympathetically, the 'What, Why and How' of arithmetic teaching at the level under consideration. The author then suggests some topics or projects which might be undertaken, each of which calls for a measure of calculation. Since so many of the difficulties and misunderstandings experienced by pupils are peculiar to a particular pupil, the chapters on 'Some underlying concepts' and 'Learning from the past' should go some way towards helping teachers to identify such particular problems and then to take appropriate action. For example, it might help the pupil to re-see the problem from a historical viewpoint, and thus to come to grips with it from an angle unfamiliar to him.

Leaving the philosophical probings to the first part, the second part of the book introduces the use of a variety of calculating aids and substitutes. Devices which include tables, graphs, number lines, nomograms, the slide rule, logarithms and digital calculating machines are not only introduced, but their potential uses with pupils are delineated in some detail, Since pencil and paper calculations are often found to be the bugbear of otherwise reasonably competent pupils, this work on aids to calculation assumes great importance, if such pupils are not to continue to suffer a heavy handicap in their search for mathematics.

Since the acquisition of arithmetical skills plays a part, sometimes a painfully suffocating part, in the common core work of almost every school, this book is likely to be studied by many teachers, whatever selection they may make from the other books of the series. This book will help 'our pupils', to overcome, or perhaps to sidestep, one of their besetting weaknesses, and their interests and energies will then be available for more exciting and worthwhile mathematical pursuits.

Number Appreciation [NA]

This book is in the main a study of the number systems in common use, and is complementary to the book *From Counting to Calculating*. The latter book deals with number notations and the skills of calculating, whilst this book seeks to codify and systematize the number knowledge which a pupil may fairly be expected to have gathered, and to extend such knowledge further. It presents a body of material to help both teachers and pupils to see number perhaps in a different light from that in which it was originally learned, and thus to provide an opportunity for a *re-seeing* or a true *revision* of past number work.

Since the work is intended for indirect use by older secondary pupils, some previous number knowledge must be assumed. Briefly this amounts to:

(i) the place-value system and the Hindu-Arabic numeral system;
(ii) ordinal and cardinal aspects of number;

(iii) some knowledge of the operations [CC1] addition, multiplication, subtraction and division;
(iv) some knowledge of fractions both common and decimal.

The chapter headings and their contents in brief are as follows.
1. Tools of the number trade. (Sets of numbers in common use — the natural numbers, the integers, rational and irrational numbers, real and complex numbers.)
2. Systematizing and codifying actions. (A discussion of *operations* where order and grouping are significant or not significant, leading to the fundamental number laws, Commutative, Associative, Distributive). Identity elements — Inverses.
3. The pieces of a number. (Common fractions — Farey series — multi-base systems leading to decimal fractions — rounding off numbers).
4. Directed number. (Activities intended to help pupils in their study of directed number — negative numbers have a mathematical existence in their own right as well as modelling certain situations in the physical world.)
5. Classifying and patterning. (Classifying numbers, odd/even, prime/composite, square numbers, triangular numbers, cubic numbers — patterning of some series, arithmetic, geometric, Fibonacci, — exponential growth.)
6. An arithmetic of remainders. (A diverting excursion into the realm of modular arithmetic.)

Suggestions for further reading by teachers and by pupils are contained in a selective book list.

This book does not employ an axiomatic approach; the work is based on things to do. The text contains a number of suggestions on which pupil activity can be based.

Algebra of a Sort [AL]

The study of traditional algebra is often regarded by many pupils and sometimes by some teachers as the black sheep of mathematics learning. This book takes pains not to perpetuate the kind of algebra which has been largely meaningless even to more able pupils and possibly to some teachers. Meaning and purpose have been imparted to the work by making the study of formulae the basis for understanding and development, with an emphasis on the importance of helping pupils to make generalizations from a study of a number of particular cases. The world of the engineer, technician and craftsman can provide the raw materials for this study, as well as for some of the other branches of mathematics, number, geometry and so forth.

So much for the general contents. The approach advocated, and amply demonstrated, is one of inductive reasoning, in which the reader is led by stages to form a hypothesis, 'It looks as if. . .', and then to seek verification. The formal deductive approach is in the main regarded as inappropriate to this first stage of algebra study and to 'our pupils'.

The opening chapter deals with generalizations based on the study of a number of cases. This chapter is followed by a study of induction and generalization from patterns, taking its examples mainly, but not exclusively, from commonly occurring number patterns. Again the approach is one of first recognizing the pattern, then expressing the findings verbally; the findings being distilled gradually and by stages into a formula, later to be refined and manipulated as required. The final chapter deals with the understanding and use of formulae. It is here that the usual algebraic skills and techniques have a part to play, but within a context and for a recognizable purpose. Albeit the author acknowledges that not all the content of this chapter is put forward as necessary or desirable on educational grounds. In short, some topics are included because of external demands which teachers and pupils may be called on to satisfy. A few topics then are introduced with some reluctance. The final decision as to what to reject and what to use must be left to the teachers concerned.

Some Simple Functions [SF]

The central theme of this book is the study of relationship, which in some senses is a sort of master key which can unlock many doors to mathematics. Some of those doors are opened in the text, but others appear either explicitly or implicitly in many of the other books in this series. As well as freeing mathematics from irrelevant detail, functional dependence, otherwise relationships, has a strong unifying influence on the many elementary branches of mathematics. The functional approach links aspects of mathematics which as a rule seem disconnected and departmentalized, and thus many of the elementary skills of arithmetic, algebra and geometry are put to use in an intermingled fashion.

It is recognized that the mathematics courses pursued by 'our pupils' are elementary in character; but even the most elementary mathematics course should contain some element of the functional approach. Every pupil has the right to experience some work in this particular field even though at lower ability levels it may be rather limited.

Having said that, it is necessary to add that essentially this is a book for *teachers*, and as such, some of its contents go well beyond a reasonable expectation for pupil use. The book aims to provide teachers of mathematics with a background which will deepen their understanding, and give them insight into mathematical thinking. The degree to which this aim is realized will determine the value of the book to teachers of mathematics and hence to pupils in their care.

Chapter 1 of the book elucidates the concept of functionality and provides a variety of elementary functions, some commonly occurring, others more rare. Thereafter the book dealt in an elementary way with a number of particular functions which include:

> The direct proportion function
> The linear function
> The 'square law' function
> The inverse proportion function
> A view of areas and volumes
> Laws and formulae from numerical data
> Periodic functions
> An important growth function (exponential growth).

Here then is a book, wide ranging in its mathematics but with a unity born of its central theme.

Geometry for Enjoyment [GE]

This book makes suggestions for creating in the classroom a structured and more precise environment within which the pupils can work mathematically. It does this by examining a variety of materials which the pupil can pick up, move, turn over, fold, join or dissect. The materials are geometrically structured, based on polygons, circles, tessellations, cubes, or equal lengths or angles. They present problems to be solved, or situations to be investigated, unlike the real world which is a result of problems already solved and situations already explored.

The book is written from the viewpoint that the normal environment provides insufficient openings for truly mathematical study through discovery. It claims that rarely in such circumstances do we find a geometrical problem or a geometrical structure worth investigating.

The normal environment has a further disadvantage in that it cannot be handled or manipulated. We cannot pick up a house and turn it round, take out window panes and refit them in their frames, take out struts from frameworks to see when they fall apart, or fold the playground in half.

In the opening chapter the author describes some of his experiences whilst teaching mathematics to a class of fourth-year leavers. From this very down-to-earth beginning there develops a series of geometrical ideas together with a wide range of materials which pupils may use in learning about them.

The chapter headings are:

> 4S (the class of fourth year leavers)
> Topological topics
> The use of materials
> Playing with polygons
> A miscellany of materials

Some teaching notes
Some geometrical notes.

In the appendices will be found a list of commercially produced materials with names of manufacturers and/or suppliers, and a list of books on the kind of geometry dealt with in the text.

Machines Mechanisms and Mathematics [MM]

This book is a stimulating survey of many forms of simple mechanisms which we all meet in our everyday life together with a discussion of the mathematical principles underlying them. From relatively common artefacts, a wealth of mathematical material emerges. The book is written in the belief that much of mathematics has developed in response to the need to solve practical problems, and that one motivation for doing mathematics will be to study its manifestations in some of its applications.

Whether we look at the gears on a bicycle, the behaviour of a car jack, the motion of a rocking horse or the design of a washing machine we are dealing with objects which have been designed to fulfil a specific purpose. The shapes of the objects and the ways in which their parts are related are highly significant, and it is the author's belief that the study of these provides a realistic alternative to traditional geometry, and that such a study can be both creative and rewarding to girls as well as to boys.

The main part of the book contains a great deal of material, but it is in no way intended to be exhaustive. The first chapter on 'The work and its presentation' suggests ways of using the material, and should be read in conjunction with the other chapters. Subsequent chapters deal in turn with:

> Linkages
> Transmitting rotary motion
> Rotary and linear motion combined
> Shapes for a purpose
> Loci

Even such a bald listing of headings gives an inkling of the variety of mathematical topics waiting to be discovered (under some guidance perhaps) from a study of everyday things.

At every stage the text is liberally supported with pupil assignments which can be used directly, or with minor amendments, by most of the pupils under consideration. There are 66 such assignments.

Here then is a lively book differing in flavour from most mathematics books, full of interesting material. It presents challenges to both teachers and pupils, all arising from some observable objects in our and their everyday worlds.

Luck and Judgement [LJ]. A classroom approach to probability and statistics

A study of probability and statistics, in one form or another, is rapidly becoming a pleasing feature of school courses in mathematics. This book seeks to introduce this topic to pupils, basing the work on experiments and activities conducted by the pupils themselves.

Following a preliminary discussion are 32 such activities and experiments presented without comment. The mathematics arising from each of these experiments is examined in detail in later chapters. Chapter 4 analyses the first 21 activities which are all *a priori* situations, that is to say it is possible to achieve a theoretical measure of probability by considering all of the possible outcomes.

An analysis of the remaining 11 activities is given in chapter 5. This set is *not* based on *a priori* situations, in other words it is *not* possible to calculate a theoretical probability by analysing all possible outcomes. In these cases, any measure of probability is a *statistical* one, based on past events or on present conditions.

Part One of this book is therefore based entirely on this selection of activities and experiments. It should be noted that all the apparatus required is of a simple nature, and is readily obtainable.

Part Two of the book begins the task of bridging the gap between the experimental, experiental work of Part One and the mass of knowledge and practice found in the real world and in more formal texts. Among the topics coming under consideration will be found:

> Arrangement and choice (permutations and combinations)
> Questionnaires and information storage (punched card systems, notched cards, coincidence cards, etc.)
> Sampling (including further classroom experiments)
> Calculating probabilities
> Comparing statistics (measures of central tendency, measures of dispersion, some special distributions).

An appendix gives a list of further reading both for teachers and for pupils. Although much of the book, Part One particularly, is written in direct form, a further appendix makes some suggestions for presenting instructions to pupils.

Here then is a book based on activities aimed at providing the pupil with a practical basis for classroom discussions which are themselves the main vehicle by which he may come to appreciate some of the ideas of probability and statistics, especially as they are likely to affect him.

Mathematics from Outdoors [MO]

An increasing number of teachers look for mathematical ideas outside the classroom, and away from printed texts. This book concerns itself mainly with simple surveying, navigation and related matters and thus should help teachers in their search for mathematics in action. Whilst there is a great deal more outdoor mathematics than is covered by this book, the topics selected are dealt with in detail, and with mathematical principles and concepts well to the fore. Although the situations posed are environmental in nature, the book stresses the mathematics inherent in such situations, and it notes the importance of intuition and discovery in the learning process.

In presenting the general principles in chapter 1, the author states: 'Two of the challenges with which a teacher is always faced are, first to provide incentives for his pupils which will give them an interest in their work and make them feel that it is worthwhile and purposeful; and secondly to ensure that they are acquiring ideas, concepts and skills which have educational significance.' This book seeks to meet these two challenges simultaneously.

After enunciating some general principles, the book considers successively:

> Fixing position (1)
> Uses of triangles
> Special triangles
> Circles and other interesting curves
> Fixing position (2)
> Similarity and trigonometry
> More about surveying
> Some practical considerations.

That the physical modelling of some mathematical abstractions is good teaching practice is clearly the basis of much of the work, for the book not only gives sources of simple and effective apparatus, but also gives precise hints for the construction and use of many pieces of apparatus.

The contents of this book oscillate purposefully between mathematical concepts and ideas and some of their physical manifestations which are accessible to both teacher and pupil.

Space Travel and Mathematics Volumes 1 and 2 [ST(1)] and [ST(2)]

These two books are considered together here, as the second book is a sequel to the first; the only example in the series where the order in which the books are used is of importance.

The development of space technology is a modern development which, in one way or another, must concern us all as citizens. Although much of the mathematics involved is complex, yet the basic ideas are within the comprehension of the lay mind. These are books on the basic ideas and on the simpler mathematics used.

Very few of us can or will experience travel in space at very high speeds, and under conditions of varying gravitational forces; for this aspect we must rely entirely on the reported experiences of astronauts. On the experimental side we must be content with terrestrial simulation of celestial conditons. These books are thus a mixture of narrative and of simple experiments capable of being performed by most pupils under normal conditions and with simple apparatus.

Book 1 [ST(1)]

This provides an elementary introduction to the ideas and concepts which are developed to higher stages in Volume 2. The list of chapter headings will indicate the nature of the topics introduced.

1. Introduction. (Which makes a case for such a study and includes some suggestions to teachers on how to present it to pupils.)
2. A journey to the moon. (Narrative.)
3. The men behind the rockets. (Biographical.)
4. Moving in space. (The rocket principle — velocity — momentum force — conservation of momentum.)
5. Some mathematics. (Gravity — equations of motion.)
6. A biographical interlude. (Kepler, Newton.)
7. Scale. (The solar system — a space timetable — speeds.)
8. Large numbers. (The index notation.)
9. Curves. (The conic sections — orbits of various shapes — escape speeds.)
10. There and back again. (Re-entry problems — interplanetary travel — space stations.)

As a guide to further reading, there is a book list. An equipment list is added also.

Book 2 [ST(2)]

As well as pursuing some other topics, this book discusses at somewhat greater depth, the ideas introduced in the plain and easy introduction provided by Volume 1. The list of topics therefore may appear to duplicate some of those in the earlier book. Such duplication is however more apparent than real, for the treatment in the second book constitute a continuation and not a repetition of the earlier work.

There now follows a synopsis of the contents of Volume 2.

1. Introduction.
2. Some ideas investigated. (Gravitation — escape speeds — weightlessness and acceleration — deceleration.)
3. Graphs and relations. (Conversion graphs — graphs of an earth/moon journey — direct variation — inverse variation.)

4. More about graphs and relations. (Mass and acceleration — speed/distance — inverse square — time of orbit and distance from the sun.)
5. Some mathematics. (Escape speeds — orbiting speeds — body falling under gravity.)
6. Speed. (Speeds of familiar objects — average speed — measuring acceleration.)
7. Keeping track. (Position on the earth's surface — position in three-dimensional space — navigating in space by dead reckoning, using the planets, using the stars — tracking the spacecraft — the accelerometer — measuring speeds — vectors.)

The final three chapters constitute an experimental introduction to the study of electronic computers.

8. Electronic computers (1). (Some uses of a computer.)
9. Electronic computers (2). (Computer models in cardboard — circuits.)
10. Electronic computers (3). (Flow diagrams — computer arithmetic — various computer methods.)

As in the first book, the text is supplemented by book and equipment lists.

These two books taken together provide a wealth of mathematical material for teacher and pupil study at varied ability levels. In many instances, the task of the teacher in transforming the material into a form more digestible by the pupil is assisted by the numerous workcards, worksheets and even a workbook which appear in the text.

Mathematical Pattern [MP]

This book is a background guide for teachers. It brings together, from a variety of branches of mathematics, topics in which an element of pattern is strongly emphasized. In some cases ideas of pattern are suggested, in other cases ideas of pattern are followed, but in one way or another, pattern is the essential ingredient of this book.

The book seeks to help teachers in their task of encouraging an appreciation of pattern, and of making a constant appeal to pattern in the mathematical work of their pupils. If a slogan had to be written, it might be, 'Where there is pattern, there is mathematics'. Much of the pattern and some of the mathematics could be made accessible to pupils of widely differing abilities, but that task is left to the professional skill and judgement of the teachers.

After a preliminary discussion of pattern's pervasiveness, the book deals at greater length with pattern in various contexts; for example:

(a) some patterns in numbers;

(b) pattern in shapes;
(c) pattern in graphical representation;
(d) pattern in statistics, probability and chance;
(e) pattern in number structure.

Such a bald listing of the content must fail to convey the flavour of the work or its essential readability. These can only be discovered by the reader who might like to start by beginning with any chapter which holds particular interest for him. There is no need to read the book through consecutively.

There are two appendices, the one dealing with the golden section and the other with perspective. These appendices amplify points raised earlier in the text.

In a background, or 'weaving', guide such as this, there will be found many links with a number of the other books of the series, more particularly those dealing with specific mathematical topics. This book highlights patterns which may receive only passing reference in the other books. In one sense then this book acts as a strengthener or girder to many of the other books of the series.

Crossing Subject Boundaries [CB]

In a growing number of secondary schools the traditional division of the curriculum and the timetable into clearly defined subjects is disappearing. In the midst of this regrouping, mathematics tends to retain its identity as a 'subject', while at the same time its own subject matter is undergoing continuous and substantial revision. This tendency could result in mathematics teachers remaining isolated at a time when their colleagues are learning to collaborate in their teaching. Credence could thus unwittingly be given to the view of some pupils that mathematics is an esoteric, mystical study, fit for the few, but not for the majority. This book aims to examine ways in which mathematics teachers might work together with teachers of other subjects, to their mutual benefit, and therefore to the benefit of their pupils.

The book discusses manifestations of mathematics and of mathematical thinking over a wide range of topics many of which already find a place in the school curriculum. Although the range is wide, it is by no means exhaustive, and it represents a selection of the possibilities open to teachers and to pupils alike. To suggest opportunities for working together is one thing; to translate them into practice is quite another. Presupposing that the will to work together is there, it may prove necessary to rethink the organization of the timetable and to redeploy the teaching force available. These are major considerations to be resolved by each school according to its own circumstances, when and if such a need arises.

Some idea of the topics discussed can be gathered from this list of chapter headings.

1. Introduction	8. Handicrafts
2. Working together	9. Art
3. Geography	10. Music
4. Orienteering	11. Sport
5. History	12. Hobbies
6. Environmental studies	13. Conclusion
7. Science	

An extensive bibliography is added as an appendix.

This is a book very much of our time and for 'our pupils'. It makes the point, *inter alia*, that there is no teacher of any subject in a secondary school who has nothing to contribute to a liberally conceived mathematics course or whose work might not benefit from his consulting with a mathematics colleague. Conversely, the mathematics teacher might well gain a wider view of the applications of mathematics than he held previously. In no way would the mathematics teacher relinquish his specialist knowledge; it would be put at the disposal of his colleagues and other members of his school. At the same time he would draw some of his applications from the studies of his colleagues.

Mathematical Experience [ME]

This book, which chronologically was the first of the series to be written, consists of a collection of articles by different authors.

Part One contains an expression of philosophy and approach relevant to the learning of mathematics by the pupil of average and below-average ability. The chapters of Part One are headed:

1. Why? What? How?
2. Mathematics for the less gifted pupils.

While Part One was written by members of the Project staff, Part Two consists of a number of case histories written by teachers directly concerned with work in the classroom. In them, some pioneering teachers relate how in one way or another they have tried to make the mathematical lives of their older pupils more bearable, relevant and attractive. They describe their difficulties as well as their successes. Some of the practices described in Part Two reflect some of the ideas put forward in Part One, but this is a matter of selection and not of collaborative writing. Part Two is a historical record of work done by adventurous pioneers during a particular period.

This book is a bridge between *Mathematics for the Majority Working Paper 14* (HMSO for the Schools Council 1967) and the rest of the books in the series.

The book should serve at least three useful purposes. They are:
1. as a source of support and encouragement to those teachers undertaking or contemplating changes in the mathematics curriculum;
2. to remove some of the sense of isolation which exists among many like-minded workers in the field;
3. to cause some comparison to be made of the action of Part Two with the thoughts of Part One.

Assignment Systems [AS]

It has always been a plank in the Project's platform to encourage small-group working in secondary schools. Tangible support for this outlook and approach takes the form of the book *Assignment Systems*. The essential theme of the book is an examination of ways of promoting a close personal involvement of a pupil with his mathematics learning, and indeed with his social and more general education.

In discussing aspects of assignment systems, the book deals with some objectives both general and specific, and it makes a case for *variety* in presentation designed to meet particular needs. Such variety might appear in the ways an assignment is communicated to a pupil, or in the organization of the groups or indeed in the philosophic basis on which the groups are initially formed.

Questions as to whether or not assignments should consist of a series of directions to be followed by the pupil, or whether they should be 'open-ended', are discussed; but as always, decisions must be left to the teachers concerned. In an appendix to Part One is reported part of a research study on *The Reactions of Children to Different Forms of Written Instruction*, and in this, open-ended questioning plays a significant part.

Discussion of the case in general is followed by discussion in particular, taking *Assignment Cards* as the basis. This section is illustrated by 46 cards composed by teachers to come under one, or more than one, of the following headings:

(a) *Situations dealt with*
   (i) stemming from the pupil's environment (whether natural or man-made);
   (ii) stemming from a pupil's past experience.
(b) *Type of assignments*
   (i) a long-term project, possibly involving a sequence of assignment cards;
   (ii) a relatively short-term assignment (questions involving school work and homework: individual working and group working are relevant here).

(c) *Form of presentation of the assignment*
   (i) a clear cut, wholly or partially directed format, for example, the usual kind of multi-facet assignment card;
   (ii) an open-ended format.

Readers may care to refer to the specimen cards in Part Two while they are reading Part One, for the cards illustrate many of the points made in the text.

This book makes out the case for the important part that the use of assignment systems can play at this level of teaching, and it illustrates the case by a detailed discussion of methods of preparing assignment *cards*.

A short index

Having set down a synopsis of each book in a condensed form, the general contents are now presented in the form of a short index. A selection of topics is taken and indications are given as to where references to them may be found in the several books.

For the code used in the references see pp. 3—4.

Topic   *References*

Number

Topic	References
calculating aids	[NA 4] [CC 5] [CC Part 2]
directed number	[NA 1] [NA 4] [MP 6]
direct proportion	[SF 2] [ST (2) 3] [CB 7]
exponential growth	[NA 5] [SF 9]
Farey series	[NA 3]
Fibonacci series	[NA 5] [MP 2]
figurate numbers	[MP 2] [NA 5] [AL 2]
fractions	[NA 3] [CC 4]
identity elements	[NA 1]
index notation	[AL 1] [CC 10] [ST (1) 8] [CB 7]
inverses	[NA 1]
inverse proportion	[SF 5] [ST (2) 3]
logarithms	[CC 10]
modular arithmetic	[NA 6]
multi-base systems	[NA 3] [CC 4] [CC App.2]
nomograms	[CC 8] [NA 4]
number lines	[CC 7] [NA 1]
number systems	[NA 1]
percentages	[LJ 10] [CB 7]
rounding off numbers	[NA 3]
simple series and number patterns	[NA 5] [MP 2] [AL 1] [AL 2]

slide rule	[CC 9]
squares and square roots	[NA 5] [MO 4] [SF 4]
the four operations in arithmetic	[NA 2] [CC 4] [CC 5] [MP 6]

Spatial studies

angles of elevation	[MO 3]
bearings	[MO 2] [ST (2) 7]
congruence	[MO 3]
curves	[MP 3] [MO 5] [MM 5] [SF 4] [ST (1) 9]
2-D shapes	[MP 3] [MM 5] [MO 5] [GE 2] [GE 4] [GE 5] [CB 5] [CB 9]
3-D shapes	[MP 3] [MM 5] [GE 2] [GE 5] [CB 8] [CB 9]
envelopes (including curve stitching)	[MO 5] [SF 4] [SF 5] [MP 3]
fixing position	[MO 2] [MO 6] [ST (2) 7]
golden section	[MP 2] [MP App. A]
linkages	[MM 2]
loci	[MM 6] [MP 3]
nets of solids	[GE 2]
polygons	[MP 3] [MO 3] [MM 2] [GE 4]
Pythagorean relation	[MO 4] [AL 2] [MP 2]
shapes for a purpose	[MM 5]
similarity	[MP 3] [MO 7] [CB 7]
symmetry	[MP 3] [GE 4] [GE 7] [CB 9]
topological topics	[GE 2]
transformations	[GE 7] [CB 8]
triangles	[MP 3] [MO 3] [MO 4] [MM 2] [GE 4]
trigonometry	[MO 7] [SF 8] [CC 7] [ST (2) 7]

Algebra

construction and use of formulae	[AL 3] [SF 7]
equations	[AL 1] [AL 3]
literal representation of variables	[AL 2]

Probability and statistics

combinations	[LJ 6]
cumulative frequency	[LJ 10]
frequency distribution	[MP 4] [LJ 11]
measures of central tendency	[LJ 10]
measures of dispersion	[LJ 11]
permutations	[LJ 6] [CB 12]

sampling	[LJ 8]
scatter diagrams	[MP 4] [LJ 5] [CB 3]

General topics

computers	[ST (2) 8] [ST (2) 9]
flow diagrams	[CC App. 4] [ST (2) 10] [CB 3] [CB 6]
golden section	[MP 2]
gravitation	[ST (1) 5] [ST (2) 2] [ST (2) 5]
information storage and retrieval	[LJ 7] [CB 3]
mass/weight	[ST (1) 4] [ST (2) 2]
momentum	[ST (1) 4]
motion studies	[MM 2] [MM 3] [MM 4] [ST (1) 5] [ST (1) 7] [ST (2) 4] [ST (2) 5] [ST (2) 6] [CB 4]
orbits	[ST (2) 4] [ST (2) 5]
perspective	[MP 3]
topics for pupil investigation	[ME 3] [ME 5] [ME 6] [AS Part 2] [LJ 2] [MM all] [CC 3] [CB 8]

Graphical representation in various forms and for various purposes

(Used extensively throughout the guides, but some particular references are given.)	[LJ 3] [SF throughout the book] [CC 6] [ST (2) 3] [ST (2) 4] [CB 3] [MP 4]

Examinations

The urge to examine even the below-average pupils is still strong. Evidence is accumulating of attempts being made in some areas to formulate and conduct special CSE examinations for those pupils not suited to the general CSE examination. Almost all of these attempts adopt a Mode III examination procedure, and they use a syllabus constructed by the school, or in some instances by a group of schools working together.

Working Paper 14 Mathematics for the Majority (HMSO for the Schools Council) discusses on p. 10 some of the effects, both positive and negative, of the CSE examination. *Inter alia* it cites the considerable merit in many of the examination syllabuses devised by the mathematics teachers on the panels of the various examining boards. Messrs Evans/Methuen have published for the Schools Council the *Schools Council Examinations Bulletin 25. CSE Mode 1 Examinations in Mathematics*. This is the report of a small working party set up by the Schools

Council, and gratitude is expressed to the Council and to the publishers for their permission to reproduce the following tables of data from that Examinations Bulletin.

It is here suggested that readers might well compare the contents of these tables with the contents of the guides given earlier in this chapter and so appreciate how the CSE syllabuses fit in with the materials in the guides. Perhaps teachers of CSE pupils will find something in the writings which will enrich their teaching.

COMMON CORE SYLLABUS

CSE Examination Boards

Topic	1	2	3	4	5	6	7	8	9	10	11	12	13	14
Number systems	•	•	•	•	•	•	•	•	•	•	•	•	•	•
four operations	•	•	•	•	•	•	•	•	•	•	•	•	•	•
scales of notation	•	•	•	•		•	•			•	•	•	•	•
irrational numbers	•	•							•			•		
index notation	•	•	•		•	•	•	•		•	•	•	•	•
logarithms	•	•	•	•		•		•	•	•	•	•	•	•
measures of quantities	•	•			•	•		•	•	•	•	•	•	
Approximations and accuracy	•	○			•	•	•	•	•	•	•	•	•	•
Averages					•	•	•	•	•	•	•	•	•	•
Percentage	•	•	•	•	•	•	•	•	•	•			•	•
Ratio						•	•	•	•	•	•	•	•	•
Prime factors	•		•	•		•						•	•	
Money	•	•			•	•		•					•	•
Mensuration in 2-D	•	•	•	•	•	•	•	•	•	•	•	•	•	•
Mensuration in 3-D	•	•	•	•	•	•	•	•	•	•		•	•	•
cone	•	•		•						•	•			
sphere	•	•	•	•										
Literal representation of arithmetical processes	•	•	•	•				•			•	•		•
Formulae	•	•		•		•		•	•	•	•	•		•
Graphs, use of	•	•	•	•		•	•	•	•	•	•	•		•
to compare systems	•	•	•							•				
gradient and intercept		•											•	•
Equations					•	•	•	•	•	•	•	•	•	•
Expansion of $(a \pm b)^2$ and $(a+b)(a-b)$	•	•				•	•	•	•			•	•	
Common factors	•	•	•			•	•					•	•	
Algebraic factors						•	•	•	•		•	•	•	•
Trinomial	•	•					•	•	•					
Grouping	•	•				•		•						
Geometrical representation of algebraic factors	•													
Graphical representation of data	•	•	•	•						•				
Variation	•													
2-D shapes	•	•	•	•				•			•	•		•
triangles	•	•	•		•	•	•	•	•	•	•	•		•
similarity	•	•	•			•	•	•	•	•	•	•		•
polygons			•	•	•	•	•	•	•	•	•	•		•
Line symmetry	•	•	•			•	•	•			•	•		•

COMMON CORE SYLLABUS – concluded.

CSE Examination Boards

Topic	1	2	3	4	5	6	7	8	9	10	11	12	13	14
Point symmetry	•	•					•	•	•		•		•	•
Use of drawing instruments	•						•					•	•	•
Scale drawing			•	•		•	•	•		•	•			•
construction of angles	•	•	•			•			•	•	•	•		•
perpendiculars	•	•				•			•	•	•			•
parallel lines	•	•	•			•			•		•			•
inscribed circle	•	•				•			•	•				•
circumcircle		•				•			•	•				•
copy an angle	•	•	•			•			•		•			
divide a line in a given ratio	•	•	•			•				•	•			
tangents	•													
Angles of elevation		•						•		•				
Angles in circles	•	•	•	•	•	•	•	•		•				
Congruence	•	•	•			•	•	•		•	•			
Locus		•	•	•		•	•		•	•		•		•
Corresponding and alternate angles			•	•		•	•	•		•	•			•
Symmetrical properties of circles	•	•	•			•	•	•		•				
Tangents		•		•	•	•	•	•			•			
Bearings	•	•	•	•	•			•	•	•				
Trigonometrical ratios	•	•	•			•	•	•						
Sine and cosine rules				•							•			
3-D shapes	•	•		•				•	•	•		•	•	•
Similar bodies	•	•				•								
Nets of solids			•		•		•							
Euler's theorem	•													
Theorem of Pythagoras	•	•	•	•	•	•	•	•	•	•	•	•	•	•

Note: in the case of the boards 4, 5 and 14 the common core is the whole syllabus.

A number of CSE examining boards offer topics as part of their mathematics examination arrangements. The following topics appeared in the various 1970 examinations.

    General mathematics.
    Calculus and coordinate geometry (four boards only).
    Civic mathematics.
    Statistics.
    Surveying.
    Navigation (four boards only).
    Mechanics.
    Geometrical drawing (one board only).
    Modern mathematics.
    History of mathematics (two boards only).

Since Civic mathematics (11 boards) and Statistics (10 boards) are among the most frequently occurring topics, analyses of these topics are now given. The tables are reproduced by permission from *Schools Council Examination Bulletin 25. CSE Mode 1 Examinations in Mathematics* (Evans/Methuen Educational 1972).

CIVIC MATHEMATICS

Topic	1	2N	2S	3	6	7	8	9	10	11	12	13
Wages	•		•	•	•	•	•		•	•	•	•
Percentage		•			•		•			•	•	•
simple interest	•		•								•	
compound interest	•	•	•			•			•	•	•	
The Budget						•						
taxes	•	•		•	•		•			•	•	•
income tax	•	•		•	•		•			•	•	•
savings	•				•	•	•	•		•	•	•
home budgets	•				•	•	•	•		•	•	•
hire purchase	•		•			•	•	•		•	•	•
life assurance			•			•						
Loans			•									
Rent						•		•	•			
Insurance						•		•	•			
Endowments	•								•	•		
Stocks and shares	•	•				•				•		•
Ready reckoners								•	•			
Trade and cash discount	•		•		•		•	•		•		•
Foreign exchange	•		•	•	•	•	•	•		•		•
Comparative costs	•		•			•						
Timetables	•				•		•			•		
Bank statements			•			•	•					•
bankruptcy			•				•			•		•
Invoices		•	•									
Appreciation and depreciation		•	•			•		•				
Scale drawing				•	•	•	•					
Graphs	•	•	•								•	•
Use of graphs			•									•
Solids of uniform												
cross-section										•		
Theorem of Pythagoras										•		
Statistics												•

STATISTICS

Topic	1	2N	2S	3	6	7	8	9	10	12	13	14
Graphical representation	•	•	•	•	•	•	•	•	•	•	•	•
Collection and												
tabulation of data	•	•	•	•	•	•	•					
Frequency distribution	•	•	•	•	•		•	•		•		•
Cumulative frequency	•	•	•		•	•		•		•		•
Normal distribution curve	•	•	•			•						
Binomial distribution						•				•		
Class intervals				•								

STATISTICS – *concluded.*

Topic	1	2N	2S	3	6	7	8	9	10	12	13	14
Time series		•										
Mean, median, mode	•	•	•	•	•	•	•	•	•	•	•	•
Weighted averages	•	•	•	•	•				•			
Moving averages	•	•						•				
Dispersion	•	•	•	•		•		•		•		
Range	•	•	•	•				•			•	•
Scatter diagram		•	•	•		•		•	•		•	
Correlation	•	•	•	•		•		•	•		•	
Sampling		•	•			•		•		•	•	
Actual practical work			•			•						
Elementary significance				•		•						
Combinations							•	•				
Permutations						•						

As in the previous table the syllabuses of two regions for one of the boards are included.

3 The mathematics curriculum

Introduction

The word curriculum is here taken to mean a totality of the factors, under the control of the school, which are involved with the education of the pupils. In its widest sense it is thus a flowing conglomerate of pupils, teachers, content, resources and approach.

A curriculum must be chosen to fulfil a purpose. In short, both general aims and particular objectives must be borne in mind when constructing one. It is therefore pertinent to consider some objectives. Pupils themselves so often rightly question *why* they should be doing a particular thing: teachers must have some answers which are acceptable to pupils as well as satisfactory to parents and to the world at large.

The mathematics policy of a school cannot exist in isolation, it is part of the wider and fuller educational policy which a school chooses to adopt. Some decisions will be the province of the mathematics department while other, and possibly greater, issues must be decided in plenary session. It is thus of importance that the Head of the Mathematics Department be eloquent in pleading his causes to non-mathematicians (if such folk really exist).

For our part we feel that we are on sound ground when we ask that, for the older pupils of average and below-average ability, their mathematical education should not be too widely divorced from the mainstream mathematics of the school. The approaches used and the resources available may be rather different, but the basic mathematics should be seen by the pupils to be related to what they have done before, and also to what they know their abler brethren are doing. If then, in the interests of continuity, we appear to be exceeding our brief, we plead justification. The education service is currently reaping some hurricanes resulting from the implementation of over-divisive past policies and we have no desire to perpetuate such.

Some objectives

(a) *To use mathematics as an instrument in the general, personal and social development of the individual pupil.*

The achievement of this objective raises a multitude of issues which constitute the back-cloth against which all the work is done. We point out at this stage that action resulting from its acceptance would differ from that aimed at producing trained mathematicians. For one thing, mathematical skills are important only when their *uses* are recognized by the pupil, and when he uses them as tools in a particular situation and not as ends in themselves, much less as 'knowledge' to be stored for hypothetical and unpredictable future use. For another thing, qualities which are not the sole province of good mathematics teaching assume much greater significance. Prominent among such qualities are adaptability, discrimination, judgement, perseverence, accuracy (of language as well as in the use of mathematical skills), clear thinking with a degree of rigour appropriate to the pupil, creativity and a development of aesthetic appreciation. Whilst such qualities should be part of the education of every trained mathematician, they play a central part in the education of the ordinary citizen, that is to say of the pupils with whom we are concerned. They are, in a very real sense, our justification for teaching mathematics to the 'ordinary' pupil.

So much for a general objective. We now come down to more particular objectives; the characters, so to speak, which play out their roles against the back-cloth of the general objective.

(b) *To achieve and maintain a critical view of existing procedures and situations, whatever their present natures; and to institute changes when and where they are appropriate*

To say that this would lead to a permanent sense of dissatisfaction is perhaps putting it a bit strongly. Nonetheless, a sense of dissatisfaction, if backed by the will and the means to remedy it, is far preferable to a state of blissful complacency.

(c) *To be conversant with, and practised in, a variety of approaches to meet the varied requirements of individual pupils*

Every pupil is of course an individual, but by the upper secondary stage, the range of personalities, temperaments, attitudes, likes, dislikes, skills and outlooks will have become greatly extended. Teachers of older pupils must thus be prepared to cope with such an extended range by having a variety of approaches and of courses readily available.

(d) *To seek out and to acquire materials, apparatus and all sorts of artefacts which lead to mathematics learning and which are conducive to mathematical thinking*

The implications of this objective should be self-evident. The 'magpie' syndrome for collecting bits and pieces is not out of place here.

(e) *To construct balanced mathematics courses where various facets of mathematics are each fairly and meaningfully represented*

The facets envisaged here are as follows.

1. *Utility*. Skills in calculation, using aids when appropriate, where purposefulness is apparent. Ability to approximate sensibly. Order of size and tolerance. Mathematics as the tool of the craftsman, engineer, scientist, geographer, artist and so forth. [CB] [MO] [ST] [MM] [CC]

2. *Cultural*. Mathematics as an ever-present feature of various cultural epochs. This not only raises historical issues such as the growth of number systems, measures in common use, or the simple practical geometry and surveying of the ancient Egyptians but also mathematics in its present day uses. For instance, probability and statistics, linear programming, navigation, orbitry and space travel, computer appreciation and the like. [CC] [LJ] [ST] [MO] [MM]

3. *Pattern and structure*. In number — in shapes — in operations — in graphical representation — in algebra. The investigation, discovery and and use of pattern as basic to the mathematician at work; thus leading to a growing awareness of certain fundamental cohering concepts which in turn lead to a study of functional dependence, the central theme which binds so much mathematics together. [MP] [NA] [AL] [SF]

4. *Language*. Mathematics as a means of communication, often transcending national boundaries and national tongues. Words — figures — symbols — diagrams and models — graphs — set language — matrices — flow charts — computer programming and so on.

    As in English, the practice of procedures, the 'grammar' of the language must be the servant not the master to the *use* of the language. It follows that every pupil must have a sound knowledge of the 'vocabulary' of mathematics as a basis for its proper use in coordinating and expressing mathematical thoughts and arguments.

    Not only is mathematical language a means of communication, but by virtue of its economy of means it can lead to further mathematical development. It does this be presenting essential facts and arguments in a small compass where their relationships can be readily seen, appreciated and extended. [NA] [CC] [AL] [GE]

5. *Aesthetic*. There is in true mathematics an element of beauty which justifies the use of the term aesthetic. Because of their limitations, 'our pupils' will probably not appreciate the intellectual beauty and elegance of some abstract mathematical proofs, but they can respond readily to visual beauty of shape, symmetries, similarities and pattern in general. Many pupils derive much pleasure and enjoyment by participating in such activities as technical and other drawing, curve stitching and model making both in two and in three dimensions. The guide *Geometry for Enjoyment* will prove of particular assistance in the development of this facet. [GE] [MP]

6. *As a way of thinking and working.* This is perhaps an all-pervading way of life rather than an identifiable objective, but since it must not be lost sight of, it is worthy of separate mention.

It does in fact emphasize that, through the thinking involved in their mathematics, our pupils can be helped to realize the general objective, previously outlined in (a), of being able to form judgements and to make decisions based on rational thought; a process which could be all-important to them as citizens later on in their lives.

'Mathematics develops patterns of thinking which are fundamental patterns of all thinking'; so write the authors of *Mathematics in Secondary Modern Schools*, a report prepared for the Mathematical Association (Bell 1959). How important then that attention should be given to ways and means of inducing pupils to *think* mathematically. We suggest that the chief factors in achieving this will be:

  (a) working according to scientific procedure, i.e. experiment — forming a hypothesis — testing the hypothesis — generalization [AL];
  (b) investigating problems (sometimes, but not always environmental ones), sifting the data, discarding the irrelevant but retaining the relevant, conducting an investigation and using logic, discrimination and judgement in appraising the outcome. In such working, the value of intuitive thought and of inductive thinking should be kept in mind, as should also the dangers of unverified intuitive thinking [LJ] [AL] [GE].

(f) *To preserve an element of continuity in the mathematical education of the pupil*

This objective may need the qualification 'as far as lies in our power'. Continuity and progression within a single school would appear to be feasible, but between schools, primary → secondary, or first school → middle school → high school, or whatever the shifts may be, the problems raised are much bigger, particularly if the jealously guarded autonomy of a school is to be respected.

As affecting the older secondary pupils, we repeat a statement made in the introduction to this chapter, namely, that the mathematics of those labelled 'less able', should not be widely divorced from the mainstream mathematics of the school. Furthermore that the work of the last year at school particularly, should have links with, and be seen to have links with, the mathematics the pupils have studied previously. The continuity can apply equally to the mathematical content and to the ways in which its study is approached.

The six objectives considered above will provide a framework for constructing courses which will coalesce into a syllabus in mathematics. We now consider some policies which, if implemented, would lead to the achievement in varying degrees perhaps, of the stated aims. At this stage the reader might well find it interesting to read or re-read the first part of [ME] (1 and 2).

Some policies

As between schools, there is great variation: variation in environment, in size, in structure, in intake and in general ethos. It would therefore follow that a school will adopt policies leading to solutions of its own particular problems. Writing for a hypothetical school can only be a generalization, so that appears to be the best we can offer in such circumstances.

Policies
1. That there be a room or rooms identifiable as the mathematics 'centre' of a school; and that such a centre, suitably furnished and equipped, shall be used by all the pupils of the school including, or perhaps especially, the less able.
2. That the average and below-average pupils get at least a fair share of specialist mathematics staff and of the resources. In some instances such a demand may only be met by an increase in the number of teachers qualified to teach mathematics.
3. That consideration be given to flexible grouping systems appropriate to given circumstances, e.g. teaching in large groups (year groups, classes), in small groups (two or three pupils) or sometime by individual work. [ME] records some experiences of these, and [AS] discusses some ways and means.
4. That the time-table shall not place limitations on work; rather it should be flexible enough to accommodate a great variety of conditions, e.g. the groupings mentioned above, or interdepartmental collaboration (see [CB]) perhaps amounting to some form of integrated studies in which mathematics is a component. The time-table must be the servant and not the master.
5. That the Head of the Mathematics Department is sufficiently freed from teaching duties to be able properly to exercise the responsibilities of the post. The Head of Department must not only display an interest in the work and progress of the average and below-average pupil, but he must be seen to display that interest in a manner recognizable by everyone, not least by the pupils concerned. This implies not only the coordination and general supervision of the work being done by pupils, but also the provision of resources and the giving of professional support and advice to members of the department, particularly those from other departments who are 'teaching some mathematics'. In any form of collaborative or integrated study, the field of such action would be extended accordingly.

6. That the 'syllabus', designed for 'our pupils' should be atomic in character. That is to say it should emerge from a large number of small, elective *short courses*, each lasting for only a few weeks at the most. This would provide for choice, in which the individual pupil, with some advice, might well have a voice in the scheme of study to be pursued by him at his own best pace. The constrictions of a common syllabus to be worked by all could thus be eliminated. This theme is developed in greater detail in the next chapter.

Whilst some of these policies, even from such a selective list, are departmental affairs, others of them can only be implemented within the context of the whole school and its organization. For some schools they might mean only readjustments, while for others, they represent major upheavals. Whichever is the case, they may be taken as *minimum requirements* for the pleasurable and efficient working of a mathematics department. Some teachers may find it necessary or desirable to add to the list such policies as are required to meet some particular local conditions.

Chapter 7 of this book, which is very short, presents some topics for discussion by teachers. The questions asked there (but not answered) are all relevant to the content of this chapter, and are intended to provide guidance in the all-important matter of decision-taking in curricular affairs.

4 Some mathematics courses

Introduction

It is well at this stage to remind the reader that the pupils who are our particular concern are those of average and below-average ability in the 13—16 age range. We have no crystal ball which will enable us to examine the past experiences or mathematics courses taken by any particular set of such pupils; neither is it within our terms of reference to suggest a syllabus to cover the whole secondary range. We can only repeat our plea that continuity throughout his secondary schooling be readily apparent to the pupil. There is a distinct possibility that the kind of content and approach which we suggest as appropriate for the latter years of schooling may have an influence on what is done, and on how it is done, in the earlier years. For those teachers who keep examination prospects in mind, a comparison of the index of the broad contents of the guides found on pp. 17—19 with the content of CSE syllabuses found on pp. 20—23 should provide some degree of comfort.

As an example of what should be done on a much larger scale we shall suggest a number of units of work which fall broadly under each of the facets found on pp. 26—27, and then combine these into courses, always aiming at a balanced total result; thus illustrating the 'atomic' principle in making a syllabus for a situation or for a pupil or a group of pupils. Further necessary extensions or modifications of the idea we must leave to those most closely concerned with the matter, the teachers in a particular school or area.

Longer term studies possibly of the environmental kind, e.g. Domestic Buildings, Communications and the like, such as are being produced in kit form by the Continuation Project, are by no means ruled out. They incorporate the various facets to varying degrees. In any case, the flexibility of the suggested procedure eases rather than inhibits the undertaking of longer-term studies worked in conjunction with the short-term ones. In point of fact, such projects may even initiate appropriate short-term studies.

Some possible short courses

Number
N 1.   Ancient numeration systems for recording numbers

(i) Simple tally systems.
(ii) Code systems with and without some element of place value.
(iii) Roman numerals — part tally — part code — some element of place value.

N 2. The Hindu/Arabic numerals. Their history including their spread throughout Europe. Their economy of means leading to algorithms or ways of calculating using numerals. Place value.

N 3. Calculating devices using only numerals for recording purposes, e.g. Abacus, Chinese Abacus, Japanese Soroban, Counting Board.

N 4. Calculations using numerals avoiding 'standard' methods. Napier's Rods; Gelosia grid for multiplication; two way entry tables including ready reckoners.

N 5. Calculations from a pair of number lines leading up to but not including the slide rule.

N 6. An introduction to nomograms.
N 7. Nomograms for specific purposes (more advanced).
N 8. Study and use of slide rule (more advanced).
N 9. 'Square' numbers and square roots.
N 10. Triangular numbers.
N 11. Other figurate numbers — generalizing into general expression.
N 12. Decimal fractions — (history and expression in terms of).
N 13. Simple common fractions.
N 14. Farey series.
N 15. Fibonacci series and its manifestations.
N 16. Pascal's triangle.
N 17. Growth by an arithmetic series
N 18. Growth by a geometric series.
N 19. Growth by a power series.
N 20. Binary number and notation.
N 21. Number systems in common use — natural numbers, integers, rational numbers, irrational numbers — leading to the idea of the real number system. Imaginary numbers in brief.
N 22. The laws of the operations of arithmetic.
N 23. *Discovering* the capabilities of a desk calculator.
N 24. Speeding up computing on a desk calculator.
N 25. Some modular arithmetic using, 12, 24, 6 and 7 as moduli.
N 26. The use of logarithms as computing aids.

Space and shapes

S 1. Informal investigations into the properties of triangles.
S 2. Informal investigations into the properties of polygons.
S 3. Modelling polyhedra.

S 4. Triangular linkages.
S 5. Quadrilateral linkages.
S 6. Tessellating plane shapes (2D).
S 7. Tessellating solid shapes (3D).
S 8. The right triangle and the Pythagorean relation.
S 9. A study of angles and rotational ideas.
S 10. The right-angle in a surveying context (Offset survey).
S 11. The right triangle in a trigonometrical context (tan, sin and cos).
S 12. Angles as bearings and as references (latitude, longitude).
S 13. A survey by triangulation.
S 14. Similar figures — scaling down — scaling up — map making.
S 15. A plane table survey.
S 16. Producing curves: (a) by graph drawing; (b) as envelopes (curve stitching/drawing); (c) as loci; (d) curves of constant width.
S 17. Angle properties of circles.
S 18. Belts and pulleys.
S 19. Gear wheels and gears.
S 20. Elementary study of plane sections of some solids — sphere — cone — cylinder — cube.
S 21. Coordinate systems — cartesian coordinates — polar coordinates — grid references — latitude and longitude. Tracking a path by given coordinate references.
S 22. Transformation by translation, by reflection, by rotation.
S 23. Symmetry — line symmetry, rotational symmetry.
S 24. A study of simple architectural shapes — arches — church windows — bridges.
S 25. Planetary orbits — the ellipse — eccentricity.
S 26. Orbits of satellites and of space vehicles.
S 27. Networks.
S 28. Unicursal routes.
S 29. Navigation charts for water/sea navigation.
S 30. Loci.

Probability and statistics

P 1. Experimental work in *a priori* situations. (*Luck and Judgement* Chapter 2 gives 21 such investigations.)
P 2. Experimental work in non-*a priori* situations. (*Luck and Judgement* Chapter 2 gives 11 such investigations.)
P 3. An information survey — collecting — presenting — storing the information.
P 4. Experimental work on sampling [LJ 8].
P 5. Competitions of the 'put in order' type.
P 6. Football pools — chances of winning in various types of pool.

P 7. Statistics from newspapers and magazines compared and discussed.
P 8. Measures of central tendency — mean — mode — median values.

A selection of more general, longer term studies
These are to be taken in conjunction with the shorter units of work. It is unlikely in general, that 'our pupils' will maintain interest in any one topic for longer than say a school term, hence it would be prudent to think of this time span in this context and to tailor the study accordingly.

G 1. Local communications — road — rail — river — telephones.
G 2. National communications — motorways — railways — canals — air travel.
G 3. Communications media — press (local and national), radio, television, journals and magazines.
G 4. Domestic buildings — designing and building.
G 5. Church/castle architecture — romanesque — gothic — modern (3D for roofs).
G 6. The mechanics of a bicycle.
G 7. The mechanics of a motor car.
G 8. A study of speeds — animals, fish, birds, men, machines, rockets. This could involve measures of timing as well as of distance.
G 9. Aeroplanes and space vehicles.
G 10. The solar system.
G 11. World-wide air services.
G 12. Study of a river from source to sea — fall in various reaches — currents — sinuosity — uses made of the river.
G 13. Study of a port — tides — facilities — volume of trade both inwards and outwards — land links with industry and centres of population. Reasons for placing the port where it is.
G 14. The mathematics of some particular hobby, e.g. fishing, model making, bell ringing, etc.
G 15. Population study of an area — total population — how distributed — some form of comparative study which might indicate trends of growth or of depopulation. Reasons?
G 16. A study of the school catchment zone — its shape — area — distribution of school population within it — journeys to and from school. Is the school well placed within the zone?
G 17. The mathematics of dress and fashion. Quantities of cloth required — mini versus maxi — design and symmetry — making a dress pattern — economical laying out of pattern on cloth length.
G 18– 44. Chapter 3 of the book *From Counting to Calculating* is devoted to some projects which involve calculating. Twenty-

six such projects are discussed there, and they include some topics commonly regarded as 'Civic Arithmetic'. The list given above is intended to supplement and not to supplant the list found in *From Counting to Calculating*.

Attention is drawn to the following book: Craddy, O. *Topics in Mathematics* (Batsford 1967).

To sum up our present position, we have listed some short units of work in the N, S, and P series and some longer term studies in the G series. No one list is intended to be exhaustive; many other possibilities are practicable, and it is hoped that teachers will extend the lists to suit the circumstances of their own situations. At this point it is well to restate that a study chosen from the G series may pinpoint the need for a shorter, more concentrated study from one of the other lists. To accommodate such an eventuality, it is essential that the organization be flexible. It is also essential that the pupils or group of pupils, play a significant part in choosing the topic, particularly from the G series; in which case it may safely be assumed that they have either a natural or a stimulated interest in the topic in question.

In terms of the material resources required, it is often advantageous to have a number of different topics running simultaneously, but the number should not place too great a strain on the resources of the teacher who has to guide and supervise the progress of the work in each of them.

Example A

Example A postulates a group of pupils who find difficulty in computation by the standard techniques, and who therefore need the support of a calculator (either a desk calculator, or better still, some form of cheap electronic calculator).

Term 1
- N 1. Numeration systems.
- N 3. Calculating devices using numerals for recording purposes.
- N 23. Discovering the capabilities of a desk calculator.
- S 1. Informal investigations into the properties of triangles.
- S 4. Triangular linkages.
- S 6. Tessellating plane shapes.
- P 1. Experimental probability work in *a priori* situations.
- G 1. Local communications (if the group is interested in this study).

OR
- G 11. World wide air services (if of interest).

OR

One other centre of expressed interest.

Term 2
N 4. Calculation avoiding the standard methods.
N 24. Speeding up computing on a desk calculator.
N 12. Decimal fractions.
N 14. Farey series.
S 2. Informal investigations into the properties of polygons.
S 5. Quadrilateral linkages.
S 9. A study of angles and rotational ideas.

Centre of interest
G 6. The mechanics of a bicycle.
OR
G 7. The mechanics of a motor car.
OR
One other centre of expressed interest.

Term 3
N 2. The Hindu/Arabic numerals — history — spread — use.
N 9. Square numbers and square roots.
N 16. Pascal's triangle.
S 8. The right triangle and the Pythagorean relation.
S 14. Similar figures — scaling up — scaling down.
S 16. Producing curves.
P 1. Experimental work in *a priori* situations (continued).
G 1. Communications media.
OR
G 14. The mathematics of some particular hobby.
OR
One other centre of expressed interest.

Example B
Example B postulates a group of pupils with reasonable computing skills either with a calculator or by standard methods.

Term 1
N 5. Calculations from a pair of number lines.
N 13. Simple common fractions.
S 17. Angle properties of circles.
S 18. Belts and pulleys.
S 21. Coordinate systems — fixing positions.
S 22. Transformations by translation, reflection, rotation.
P 5. Competitions of the 'Put in order' type.

Some possible general topics.
(a) Insurance — gambling [CC 3]

OR
(b)   The Welfare State [CC 3]
OR
(c)   A fair day's pay [CC 3]
OR
(d)   Investments [CC 3].

Term 2
N 17.   Growth by an arithmetic series.
N 18.   Growth by a geometric series.
N 19.   Growth by a power series.
S 3.   Modelling polyhedra.
S 10.   The right-angle in a surveying context (Offset survey).
S 19.   Gear wheels and gears.
S 24.   Simple architectural shapes.

A centre of interest possibly selected from:
G 16.   A study of the school catchment zone.
G 13.   A study of a port (if relevant).
G 17.   The mathematics of dress and fashion or some other topic which gains general approval.

Term 3
N 6.   An introduction to nomograms.
N 20.   Binary number scale — notation — applications.
S 7.   Tessellating solid shapes.
S 12.   Angles as bearings and as references.
S 23.   Symmetry — line symmetry, rotational symmetry.
S 20.   Elementary study of plane sections of some solids.
S 27.   Networks.

A centre of interest from:
P 3.   An information survey — collecting — presenting — storing information.
G 2.   National communications systems.
G 12.   Study of a river (if relevant).

Example C
Example C postulates a group of self confident pupils who have had previous experience of the kinds of courses outlined in examples A and B. In short, this course assumes some knowledge found in the earlier examples.

Term 1
N 7.   Nomograms for specific purposes.

N 21. Number systems in common use.
S 11. The right triangle in a trigonometrical context (tan/sin).
S 13. A survery by triangulation.
S 28. Unicursal routes.
P 2. Experimental probability work in non-*a priori* situations ([LJ 2] gives 11 such investigations).

A centre of interest possibly selected from:
G 15. Population study of a local area.
OR
Even if apparently repetitive, a deeper and more advanced study of a topic taken earlier, e.g.
   Study of a river (if relevant).
   Study of a port (if relevant).

Term 2.
N 8. The slide rule.
N 15. The Fibonacci series and its manifestations.
S 15. A plane table survey.
P 4. Experimental work on sampling [LJ 8]
S 11. The right triangle — heights and distances.

For general study:
An ecological study of one or more local habitats — collecting — presenting and storing information.
OR
A study of the National Savings Movement — its activities and possibilities.

Term 3
N 8. Speeding up and extending the use of the slide rule.
N 25. Some modular arithmetic.
N 26. The use of logarithms (if you must use them).
S 25. Planetary orbits.
S 26. Orbits of satellites and space vehicles.
S 30. Loci.
P 8. Measures of central tendency — mean, mode, median.

General study
P 7. Statistics from newspapers collected, compared and discussed.
OR
G 10. The solar system.
OR
Space travel — moon journeys.

Example D
Example D approaches the situation from a different angle. An agreed wide study becomes the *starting point*, and whatever short courses prove to be necessary or desirable emanate from the central theme. In such circumstances it is obviously of mathematical value to choose a topic which contains or suggests such further developments, rather than one in which mathematics is sparsely represented.

Let us assume that the agreed study is G 11: World-wide air services.

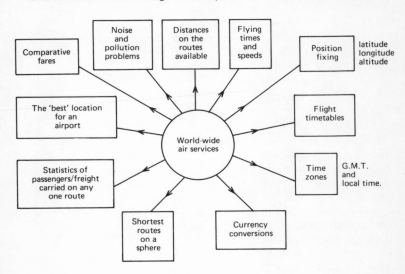

Once the general theme is introduced and discussed, any one of the topics in the outer zone could be the subject of short intensive study. Mathematical skills and techniques thus appear for what they really are, tools for doing a job.

Example E
Example E is another example of short courses springing from a central project. In *Working Paper 14* (HMSO for the Schools Council) on p. 14, reference is made to what is there called the mathematical exploitation of non-recurring situations. This project comes in that category, although one school, having successfully planned and built a pig and poultry unit, repeated the project in effect by building a calf-rearing unit; but of course the work was done by a later generation of fourth year pupils.

Motivation in this kind of work is indeed strong, as the final outcome is there for all to see, and for generations of pupils to use and to admire.

The project was the *planning and building of a pig and poultry unit* for use by the Rural Studies Department.

1. The planning phase.
   (a) Decisions as to space requirements, and the best disposition of space units within the total unit.
   (b) Dimensions which ensure that wall lengths are commensurate with the building units used (bricks/concrete blocks) and that wall heights are similarly commensurate.
   (c) Internal walls to tie the building together leading to a discussion of stability and rigidity.
   (d) The weight of a brick/concrete block — the weight of the structure to be supported — pressure (lb/ft^2 or kg/m^2) to be carried by the foundations.
   (e) The nature of the subsoil determining the foundations required.
   (f) Volume of trenching for the foundations or footings.
   (g) Roof trusses to be used — slope and nature of roofing.
   (h) Areas of internal and external walls to be rendered, plastered or finished in one form or another.
   (i) Other important details such as damp course, doors, windows, ventilation, rainwater disposal.

2. The modelling phase.
   Giving effect to the decisions made by drawing up detailed working drawings of plans and elevations. The mathematical content of this phase is self evident. It is a reasonable test of the working drawings if a card model of the structure is made from them at this stage. Mistakes are easier to rectify on paper than in an actual building.

3. Materials — quantities and cost estimates.
   e.g.

		£
Concrete blocks	560 x 6 in	
Ready mixed concrete for footings	5 yd^3	
Cement	20 cwt	
Graded sand	$4\frac{1}{2}$ yd^3	
Asbestos roofing	20 sheets	
Windows		
Doors		
Timber	200 ft 6 in x 2 in	
	200 ft 3 in x 2 in	
	100 ft 5 in x 1 ft	
Nails		
Gutters		
Paint	etc	
Damp proof course		
	Estimated cost	

4. The construction phase.
    (a) Assembling the materials on the site.
    (b) Setting out on the site. (A very important piece of practical mathematics involving levels as well as distances and right-angles.)
    (c) Keeping the walls 'true', both vertically and horizontally.
    (d) The carpentry of the roof trusses.
    (e) Finishing the building — rendering/plastering/painting.
5. Finally comparing the estimated quantities and costs with the actual quantities used and their costs.

Such a wide-spreading project is clearly the work of many hands and of several departments of a school and as such it may fairly be claimed to cross subject boundaries in a most meaningful way. (CB 8) refers to a like project namely the building of a garage.

Synthesis or analysis?

More often, this kind of project takes the form of a study of an existing house, in which case an analytic viewpoint is adopted. In planning houses for human habitation, criteria apply which are different from those used in the previous illustration, and this leads to more complex structures and a greater variety of forms. So whilst material demands will be different and perhaps easier to cope with, a domestic habitation is a more complex unit in form, structure and decoration. Nevertheless it can provide a valid study, as evidenced from the work of the Continuation Project in producing a kit of pupil material based on this theme.

The five examples from our chosen limited field illustrate a procedure which can be extended to cover the complete mathematics curriculum for 'our pupils'.

This chapter concludes with three important thoughts.
(a) A curriculum based entirely on wide topics of the examples D/E types, is likely to be out of balance. Even though succeeding topics are chosen with care, it is unlikely that each of the facets mentioned earlier will be fairly represented — indeed some could be omitted altogether. It might well be thought that such wider studies are more suited to fifth year working. The 'atomic' or 'patch' approach illustrated by examples A, B and C is likely to provide a more balanced syllabus, as well as ensuring variety in the work.
(b) The virtues of 'concentric' study should not be overlooked. The fact that a pupil has previously worked a topic perhaps in the junior school, or in the lower secondary school is of itself not a reason for excluding it from the upper school. The services of the Post Office could serve as an example of a theme which can be worked at very varied levels of sophistication. The same might be said of the ever popular curve stitching — leading to the production of envelopes by

drawing and hence to the geometrical properties on which the construction is based. Provided that there is an increased challenge to skills, and an extension of ideas and vocabulary, there is much to commend concentric working. Mere repetition would be not only devastatingly dull, but would defeat its own ends, and must be avoided at all costs.

c) It is often said that the operation of curricula on the lines suggested would prejudice the eventual chances of a late developer. Such fears can be discounted if: (i) it is realized to what extent the subject matter covered by the Project's guides in reflected in CSE syllabuses, as is shown in chapter two of this book; and (ii) full advantage is taken of a concentric (or spiral) development, such as that just referred to, when the late developer and/or a less-able companion can be studying the same topic together at quite different mathematical levels.

5 A book list

This list is divided into two sections. Section A represents books more suited for use by teachers whilst Section B gives some books which may be used by pupils. The dividing line is of course arbitrary and rather ill defined.

Section A. Books for teachers set out by sections

A1. Books on number

Adler, I., *The New Mathematics* (Dobson).
Baker, C. C. T., *Introduction to Mathematics* (Newnes Butterworth).
Bowers, H. & Bowers, J., *Arithmetical Excursions* (Dent).
Brodetsky, L. S., *A First Course in Nomography* (Bell, 1920).
Colerus, E., *From Simple Numbers to the Calculus* (Heinemann).
Combridge, J. T., (Ed), *Count Me In* (Queen Anne Press).
Dantzig, L. T., *Number, the Language of Science* (Allen and Unwin).
Dienes, Z. P., *Building up Mathematics* (Hutchinson).
Eves, H. W. & Newson, C. V., *An Introduction to the Foundations and Fundamental Concepts of Mathematics* (Holt Reinhart and Winston).
Gattegno, C., *Mathematics with Numbers in Colour,* Books 5, 6 & 7, (Educational Explorers).
Hamill, C. & Marshall, J. J., *Modern Mathematics* (Hulton).
Holt, M. J. & McIntosh, A. J., *The Scope of Mathematics* (O.U.P.).
Klose, O. M., *The Number Systems and Operations of Arithmetic* (Pergamon Press).
Lewis, W. D., *Teaching School Mathematics with the Desk Calculator* (Heinemann 1966).
Mansfield, D. E. & Thompson, D., *Mathematics: A New Approach* (Chatto & Windus).
Marjoram, D. T., *Modern Mathematics in Secondary Schools* (Pergamon Press).
Peter, R., *Playing with Infinity* (Bell).
Reichmann, W. J., *The Fascination of Number* (Methuen).
Reid, C., *From Zero to Infinity* (Routledge & Kegan Paul).
Smeltzer, D., *Man and Number* (Black).
Thomas, L. & Thomas, A., *Mathematics by Calculating Machine* (Cassell 1962).

Weekes, S. P., *Exercises in Arithmetic by Calculating Machine* (C.U.P.). (1966).

Yoshino, Y., *The Japanese Abacus Explained* (Dover Books).

A2. Historical

ATM, *Mathematics History — a Select Bibliography* (An ATM pamphlet).
Ball, W. R., *A Short History of Mathematics* (Dover Books).
Boyer, C. B., *A History of Mathematics* (John Wiley).
Eves, H., *An Introduction to the History of Mathematics* (Holt Reinhart and Winston).
Gartmann, H., *The Men behind the Space Rockets* (Weidenfeld & Nicholson 1955).
Midonick, H., *The Treasury of Mathematics* (2 Vols) (Pelican Books).
Pullan, H. D., *The History of the Abacus* (Hutchinson 1968).
Sanford, V., *A Short History of Mathematics* (Harrap).
Struik, D. J., *A Concise History of Mathematics* (Dover).
Yeldman, F. A., *The Story of Reckoning in the Middle Ages* (Harrap).
Yeldman, F. A., *The Teaching of Arithmetic through 400 Years (1535–1935)* (Harrap).

A3. Spatial studies

Bell, R. C., *Board and Table Games* Vols. I and II (O.U.P.).
Bell, R. C., *Tangram Teasers* (Corbitt & Hunter).
Bell, A. & Fletcher, T., *Symmetry Groups* (A.T.M.).
Coxeter, H. S. M., *Introduction to Geometry* (Wiley).
Cundy, M. & Rollett, A. P., *Mathematical Models* (O.U.P.).
Curtin, W. G. & Lane, R. F., *Concise Practical Surveying* (E.U.P.).
Dean Lent, *Analysis and Design of Mechanisms* (Prentice Hall).
Deverson, H. J., *The Map that came to Life* (O.U.P.).
Disley, J., *Orienteering* (Faber and Faber).
Gardner, A. C., *Teach yourself Navigation* (E.U.P.).
Gardner, M., *Mathematical Puzzles and Diversions* (Penguin).
Gardner, M., *More Mathematical Puzzles and Diversions* (Penguin).
Garnier, B. J., *Practical Work in Geography* (Arnold).
Golomb, S. W., *Polyominoes* (George Allen and Unwin).
Higgins, A. L., *Elementary Surveying* (Longmans).
Holmes, L., (Ed.)., *Odhams New Motor Manual* (Odhams).
Jeger, M., *Transformation Geometry* (George Allen and Unwin).
Kline, M., *Mathematics in Western Culture* (Allen & Unwin).
Land, F. W., *The Language of Mathematics* (Murray).
Levy, L., *Modern Geometry & Trigonometry* (Longmans).
Life Science Library, *Machines*.
Lockwood, E. H., *A Book of Curves* (C.U.P.).
Northrop, E. P., *Riddles in Mathematics* (Pelican).

Orienteering (Know the Game series) (Educational Productions).
Read, R. C., *Tangrams* (Dover).
Rowland, K., *Looking and Seeing* (Ginn).
Steinhaus, H., *Mathematical Snapshots* (O.U.P.).
Tahta, D. G., *Pegboard Games* (A.T.M.).
Thompson, D'Arcy, *On Growth and Form* Abridged paper-back edition (C.U.P.).
Wells, A. F., *The Third Dimension in Chemistry* (O.U.P.)
Wenninger, M. J., *Polyhedron Models* (C.U.P.).
Weyl, H., *Symmetry* (Princeton/O.U.P.).

A4. Probability and statistics

Abstract of British Historical Statistics (C.U.P.).
Adler, I., *Probability and Statistics for Everyman* (Dobson).
Annual Abstract of Statistics (H.M.S.O.).
Battersby, A., *Mathematics in Management* (Pelican).
D'Arcy, J., *Chance and Choice* (Thames & Hudson).
Glasgow Schools Statistics Panel, *Statistics and Probability* (Foulsham).
Griffiths, S. & Downes, L., *Educational Statistics for Beginners* (Methuen)
Hogben, L., *Chance and Choice by Cardpack and Chessboard* (Parrish).
Huff, D., *How to Lie with Statistics* (Gollanz).
Huff, D., *How to Take a Chance* (Pelican).
King, A. C. & Read, C. B., *Pathways to Probability* (Holt).
Lockwood, E. H., *Statistics: The How and Why* (Murray).
Loveday, R., *Statistics – a First Course* (C.U.P.).
Loveday, R., *Statistics – a Second Course* (C.U.P.).
Moroney, M. J., *Facts from Figures* (Pelican).
Nuffield Mathematics Project, *Probability and Statistics* (Chambers/Murray).
Ormell, C. P., *Probability and Statistics* (Oliver and Boyd).
Reichmann, W. J., *The Use and Abuse of Statistics* (Pelican).
Sherlock, A. J., *Probability and Statistics* (Arnold).
Vessello, I. R., *How to Read Statistics* (Harrap).
Weaver, W., *Lady Luck, the Theory of Probability* (Heinemann).
Wolff, P., *Breakthroughs in Mathematics* (Signet Science Library).

A5. Crossing Subject Boundaries. A selective list of books from some other fields which provide opportunities for mathematical work.

Geography

Birch, T. W., *Maps* (O.U.P.).
Chorley, R. J. & Haggett, P., *Frontiers in Geographical Teaching* (Methuen).
Crone, G. R., *Maps and their Makers* (Hutchinson).
Hinks, *Maps and Survey* (C.U.P.).

Walford, R. A., *Games in Geography* (Longmans).
Wheeler, K. S., *Geography in the Field* (Formerly published as *Geographical Fieldwork*) (Blond Educational).

History

Atkinson, R. J. C., *Stonehenge* (Penguin).
Ellacott, S. E., *Spearman to Minuteman* (Abelard-Schuman).
Ferguson, S., *Projects in History* (Batsford).
Schneider, H., *Everyday Machines* (Brockhampton Press).
Sellman, R. R., *Mediaeval English Warfare* (Methuen).
Storm, M., *Urban Growth in Britain* (O.U.P.).

Environmental studies

Bolger, F. J., *Rural Studies* (Allman).
Hopkins, M. F. S., *Learning through the Environment* (Longmans).
Masterson, T. H., *Environmental Studies* (Oliver and Boyd).
Rigg, J. B., *A Textbook of Environmental Study* (Constable).

Science

Bondi, H., *The Universe at Large* (Heinemann).
Cohen, I. B., *The Birth of a New Physics* (Heinemann).
Feather, N., *Mass, Length and Time* (Pelican 1965).
Hilton, A. C. & Hilton, D. A., *Projects in Biology* (Batsford).
Nuffield Chemistry Series (Longmans/Penguin).
Nuffield Physics Series (Longmans/Penguin).
Nuffield Secondary Science Project, *Nuffield Secondary Science Themes 1—8* (Longmans).

Design, art and handicraft

Arnheim, R., *Art and Visual Perception* (Faber).
Building Research Station, *Project Network Analysis* (Digest 53). (H.M.S.O.).
Gombrich, E. H., *Art and Illusion* (Phaidon).
Kepes, G., *Module, Symmetry, Proportion* (Studio Vista, London).
Klee, P., *Notebooks*, Vol. *1 The Thinking Eye* (Lund Humphries, London).
Ministry of Public Building & Works, *Setting out on Site* (Advisory Leaflet 48) (H.M.S.O.).
Munari, B, *Discovery of the Square* (Alec Tiranti, Ltd., London).
Schools Council Curriculum Bulletin 4, *Home Economics Teaching* (Evans/Methuen).

Music

Mandell, M. & Wood, R. E., *Make your own Musical Instruments* (Bailey Bros. and Swinfen).
Paynter, J. & Ashton, P., *Sound and Silence* (C.U.P.).
Roberts, R., *Musical Instruments Made to be Played* (Dryad).

Sport and physical education

Amateur Athletic Association Handbook
Dyson, G. H. G., *The Mechanics of Athletics* (U.L.P.).
Jodey, J. M., *Sportsmaths* (Blond Educational).
Rothman's Football Yearbook (Queen Anne Press).
Wisden Cricketers' Almanack (Sporting Handbooks Ltd.).

Hobbies

Aeromodeller Annual (Model and Allied Publications, Ltd.).
Camp, J., *Discovering Bells and Bell-ringing* (Shire Publications).
Davis, A., *Tackle Motor-cycle Sport this Way* (Stanley Paul).
De Mare, E., *Photography* (Penguin).
Snowdon, J., *Ropesight* (Whitehead and Miller, Leeds, 1955).
Sussman, A., *The Amateur Photographer's Handbook* (Pitman 1958).
Wallace, C., *The Junior Photographer* (Evans).
World Motor-Racing and Rallying Handbook (Mayflower Books).

The facts of space travel

Alexander, T., *Project Apollo* (Frederick Muller Ltd., London 1964).
Booker, P., Frewer & Pardoe, *Project Apollo* (Chatto and Windus).
Coombs, C., *Project Apollo* (Bell).
Peter Fairley's *Space Annual.* (TV Times).
The How and Why Book of Planets and Interplanetary Travel. A 'How and Why' book (Transworld Publishers Ltd).

The mathematics of space travel

Department of Education and Science, *Learning about Space* D.E.S. Education Pamphlet Number 55: (H.M.S.O. London, 1969).
Ministry of Defence, *An Introduction to Space* (H.M.S.O., London, 196
Smith, S. W. (Ed.) *A Handbook of Astronautics* (U.L.P. 2nd Edition, 1966).

Computers

Andrew, A., *Brains and Computers* (Harrap, 1963).
Bolt, A. B. & Wardle, M. E., *Communicating with a Computer*. S.M.P. Handbook (C.U.P., 1970).
Clark, John O. E., *Computers at Work* (Paul Hamlyn, 1969).

Hollingdale, S. H. & Tootill, G. C., *Electronic Computers* (Penguin Books, 1965).
Nuffield Mathematics Teaching Project, *Computers and Young Children* (Chambers/Murray).
Nuffield Mathematics Teaching Project, *Logic and Computers* (Chambers/Murray).
Piper, R., *The Story of Computers* (Brockhampton Press).
Schools Mathematics Project, *We Built our own Computers*. S.M.P. (C.U.P.).

A6. General

An Experiment in Team Teaching, Themes in Education No. 6 (Exeter Institute of Education 1968).
Boothman, D. B., *Topical* (Longmans 1967).
Craddy, O., *Topics in Mathematics* (Batsford 1967).
Fletcher, T. J. (Ed.), *Some Lessons in Mathematics* (C.U.P.).
Glenn, J. A. & Robinson, J., *The Mathematics of Money* (Decimal Currency Ed) FREE from the National Savings Committee, Alexandra House, Kingsway, London W.C.2.
Kaner, P., *Modern World Mathematics* (5 books) (Longmans).
Kline, Morris, *Mathematics – A Cultural Approach* (Addison Wesley 1962).
Mathematics Laboratories in Schools (Bell for the Mathematical Association 1968).
Moakes, A. J., *Core of Mathematics* (Macmillan).
Moakes, A. J., Croome, P. D. & Phillips, T. C., *The Pattern and Power of Mathematics*. 7 books (Macmillan).
Paling, D. & Fox, J. L., *Elementary Mathematics. A Modern Approach* (2 books with answer) (O.U.P. 1969).
Sawyer, W. W., *Introducing Mathematics* (Penguin 1968).
Sawyer, W. W., *Mathematicians Delight* (Pelican).
Sawyer, W. W., *Prelude to Mathematics* (Pelican).
Sawyer, W. W., *Vision in Elementary Mathematics* (Pelican).
Sawyer, W. W., *A Path to Modern Mathematics* (Penguin).
Sawyer, W. W., *A Search for Pattern* (Penguin).
Sawyer, W. W. & Srawley, *Designing and Making* (Blackwell 1950).
Sobel, Max A., *Teaching General Mathematics* (Prentice Hall 1967).
Sumner, R. & Warburton F. W., *Achievement in Secondary School* (N.F.E.R. 1972).
The Mathematical Association, *Mathematics in Secondary Modern Schools* (Bell, 1959).
The New Look in Mathematics Teaching (University of Hull, Institute of Education 1965).
The Schools Council Working Paper 14, Mathematics for the Majority, H.M.S.O. 1967.

The Schools Council Curriculum Bulletin No. 1 Mathematics in Primary Schools (H.M.S.O. 3rd ed. 1969).
The Schools Council Examinations Bulletin 25 CSE Mode 1 Examinations in Mathematics (Evans/Methuen 1972).
The Schools Council, *Change for a Pound* (H.M.S.O. 1968).
The Schools Council, *Measure for Measure* (Evans/Methuen 1970).
The Schools Council, *Enquiry 1: Young School Leavers* (H.M.S.O. 1968).
The Teachers' Guides of the Nuffield (5–13) Mathematics Project (Murray/Chambers 1967 onwards).

Useful references
The banking service offers handbooks and other educational aids. Apply to your local bank branch or to: The Bank Education Service, 10 Lombard Street, London, E.C.3. which can supplement material available from bank counters.
Whitaker's Almanack (published at 13 Bedford Square, London W.C.1). Full edition £1.75 (1971). Shorter paperback edition £0.85 (1971).
The Post Office Guide (Post Office).
The A.A. Handbook (Automobile Association).
McWhirter, N. & McWhirter, R., *The Guinness Book of Records* (Doulton)
See also *catalogues* from builders' merchants, nurserymen, mail order houses, etc., as well as mathematics from the daily and the weekly press.

A7. Periodicals published in the U.K.
Mathematics in School, published six times a year by Longmans for the Mathematical Association. (Subscription rate £3 per year.)
Mathematics Teaching published quarterly by the A.T.M., Market Street Chambers, Nelson, Lancs. (75p.)
Mathematics Teachers' Forum (Ed. D. E. Mansfield) published six times a year by Fanfare Publishing House, 174 Chingford Mount Road, London E4 9BS. (Mainly concerned with primary/middle school mathematics.)

U.K. professional associations concerned with the teaching of mathematics
The Association of Teachers of Mathematics (ATM)
Headquarters: Market Street Chambers, Nelson, Lancashire BB9 7LN
Publications
    (a) *Mathematics Teaching*, published quarterly by the ATM (75p).
    (b) A long list of pamphlets on mathematics and mathematics teaching published by the ATM.
    (c) A number of books published by Cambridge University Press for the Association.
Local branches are in action in most areas of the U.K.

The Mathematical Association
Headquarters: 150 Friar Street, Reading RG1 1HE
Publications
- (a) *The Mathematical Gazette*; published by Bell four times a year.
- (b) *Mathematics in School*; published by Longman Group Ltd (Journals Division), six times a year. Subscription rate £3 per year.
- (c) A long list of reports on various aspects of school mathematics is published for the Association by Bell.

Local branches are in action in most areas of the U.K.

Section B. Some books for pupils set out by sections

In many of the guides it has been put forward strongly that a wide variety of books should be readily available for use by pupils. Such use might take various forms. For example, as a reference to some particular feature; or again, for fuller and more detailed study of a particular topic being studied by a group or by an individual. The books listed in this section include both categories. The topic study books are generally quite small and comparatively inexpensive, and might well form the nucleus of a classroom/workshop library which would be expanded as funds became available. A few books for more general reading have been included.

B1. Books on number

Adler, I., *Magic House of Numbers* (Dobson).
Adler, I, & Adler, R., *Numbers Old and New* (Dobson).
Adler, I. & Adler, R., *Numerals: New Dreams for Old Numbers* (Dobson).
Andrews, W. S., *Magic Squares and Cubes* (Dover Publications).
Bendick, J. & Levin, M., *Take a Number* (Blackie).
Fielker, D. S., *Computers* (Topics from Mathematics Series) (C.U.P.).
French, P., *Calculating Devices* (House of Grant).
Gattegno, C., *Mathematics with Numbers in Colour*, Books 5, 6 & 7 (Educational Explorers).
Hunter, J. A. M., *Figures for Fun* (Phoenix House).
Johnson D. A. & Glenn, W. H., *Fun with Mathematics*
Johnson, D. A. & Glenn, W. H., *Curves*
Johnson, D. A. & Glenn, W. H., *Calculating Devices*
Johnson, D. A. & Glenn, W. H., *Theorem of Pythagoras*
Johnson, D. A. & Glenn, W. H., *Short Cuts in Calculating*
} Topic books from the series *Exploring Mathematics on your Own* (Murray).

Mathematics for Schools — A Modern Approach (Foulsham 1967 onwards). A series of pupil texts (including Geometry).

Moss, G. A., *Think of a Number* (2 books and a teachers' book) (Blackwell).
Paling, D., Banwell, C. S. & Saunders, K. D., *Making Mathematics* (O.U.P. 1968 onwards). Pupil texts, pupil work books and topic books.

B2. Historical

Adler, I., *The Giant Colour Book of Mathematics* (Hamlyn 1961).
Adler, I., *The Golden Book of Mathematics* (Golden Press 1958).
Adler, I., *Time in your Life* (Dobson 1957).
Bendick, J., *How Much and How Many* (Brockhampton Press 1960).
Bowman, M. E., *Romance in Arithmetic* (U.L.P.).
Groom, A., *How We Weigh and Measure* (Routledge & Kegan-Paul 1961).
Groom, A., *Maps and Map Making* (Golden Press).
Hogben, L., *Man Must Measure* (Rathbone Books 1955).
Hood, P., *How Time is Measured* (O.U.P. 1955).
Rucklis, H, & Engelherdt, J., *The Story of Mathematics* (Harvey House 1963).
Smith, D. E., *Number Stories of Long Ago* (Ginn).
Smith, T., *The Story of Measurement* (Blackwell 1959).
Smith, T., *The Story of Numbers* (Blackwell).
Williams, S. E., *Stories of Mathematics* (Evans).

B3. Spatial studies

Cole, J. P. & Benyon, N. J., *New Ways in Geography* (Books 1 & 2) (Blackwell).
Escher, M., *The Graphic Work of M. Escher* (Oldbourne).
Fielker, D. S., *Cubes* (C.U.P.).
Fletcher, D. E. & Ibbotson, J., *Geometry with a Tangram* (Holmes).
French, P. *Introducing Topology* (Grant).
James, E. J., *Curve Stitching* ⎫
James, E. J., *The Aircraft Pilot* ⎬ From the series Mathematical Topics for Modern Schools (O.U.P.).
James, E. J., *Aircraft Navigation* ⎭
Johnson, D. H. & Glenn, W. H., *Topology* (John Murray).
Lewis, K., *Further Experiments in Mathematics* (Longmans).
       1. *The Ellipse.*
       2. *Polyhedra.*
       4. *Geometry without Instruments.*
Mold, J., *Circles* (C.U.P.).
Mold, J., *Solid Models* (C.U.P.).
Mold, J., *Tessellations* (C.U.P.).
Paling, D., Banwell, C. S., & Saunders, K. D., Making Mathematics: Topic book 3: *Making Models.* Topic book 6: *Creative Design* (O.U.P.)
Ravielli, A., *Adventures with Shapes* (Phoenix 1960).
Weyl, P. K., *Men, Ants and Elephants* (Phoenix 1961).

B4. Probability and statistics

Campbell, I., *Statistics* (Longmans).
Fielker, D. S., *Computers* (C.U.P.).
Fielker, D. S., *Statistics* (C.U.P.).
Fielker, D. S., *Towards Probability* (C.U.P.).
Giles, R. (Ed), *Social Statistics of Great Britain* (Clearway Books).
Giles, R. (Ed), *Transport Statistics* (Clearway Books).
Johnson D. A. & Glenn, W. H., *The World of Statistics* (Murray).
Lewis & Ward, *Starting Statistics* (Longmans).
Razzell, A. G. & Watts, K. G., *Probability* (Hart-Davis).

B5. General

Barrett, W. G. & Hollis, R. G., *Maths Packs*. Two sets of work cards. (Cassell)
Carver, C. & Stowasser, C. H., *Measuring and Making* (2 books) (O.U.P.).
Deverson, H. J. & Lampitt, R., *The Map that Came to Life* (O.U.P. 1948).
Fielker, D. S. & Mold, J., *Topics from Mathematics* (C.U.P. 1967–71). (Series of 8 titles.)
Hayman, M., *Essential Mathematics* (Macmillan).
Hogben, L., *Men, Missiles and Machines* (Rathbone Books 1957).
Hood, P., *Observing the Heavens* (O.U.P. 1951).
James, E. J., *Mathematical Topics for Modern Schools* (O.U.P.). (Series of 12 titles.)
Johnson, D. A., Glenn, W. H. et al., *Exploring Mathematics on Your Own* (Murray). (Series of 21 titles.)
Lee, L. & Lambert, D., *Man Must Move* (Rathbone Books 1960).
Lester, R. M., *The Observer's Book of Weather* (Warne 1955).
Lewis, K., *Further Experiments in Mathematics* (Longmans 1969). (Series of 4 titles.)
Midlands Mathematical Experiment, *Excursions from Mathematics* (2 books) (Harrap).
Paling, D., Banwell C. S. & Saunders, K. D., *Making Mathematics: Topic Books*. (Oxford 1971–73). (Series of 12 titles.)
Pearcey J. F. F. & Lewis, K., *Experiments in Mathematics* (Longmans). (Series of 3 books.)
Razzell, A. G. & Watts, K. G. O., *Mathematical Topics* (Rupert Hart-Davis). (Series of 6 titles.)
Whitney, T. H., *Young Scientist's Approach to the Weather* (Warne 1962).

6 Some commercially produced materials with their sources

Section A lists some commercially produced materials. Where manufacturers names only are given, the materials are normally available on local purchase, but catalogues are usually available from the manufacturers. It is usually best to obtain materials from educational suppliers. Care has been taken to ensure that the list is accurate at the time of writing but no guarantee can be given as to how long this will last.

Section B lists some films, film strips and film loops which are broadly relevant to the contents of the guides.

Section A1. Two Dimensions

Drawing

Etch-a-sketch	(Fisher, E.S.A.)
Imagidraw	(Waddington)
Reflecto-graf	(Merit)
Spirograph	(Fisher, Brightway, E.S.A.)

Pegboard games

Chinese Chequers	(Merit, Spear, E.S.A.)
Solitaire	(Merit, Spear, Arnold, E.S.A., Invicta)

Network games

Check Lines	(Triang, Invicta)
Connect	(Galt)
Troke	(Invicta)

Shapes and tessellations

Geometrical Shapes	(School Utilities)
Geometric Shapes	(Taskmaster)
Geo Strips	(Taskmaster)
Geo-metric Strips	(Invicta)
Mosaics	(E.S.A.)

Multimosaic	(E.S.A., Brightway)
Multipurpose Mosaic Shapes	(Taskmaster)
Pattern Tiles	(Taskmaster)
Pollypattens	(Invicta)
Tessellation	(Arnold)
Tessellations	(Taskmaster)

Tessellation puzzles and games

Angle	(Pan)
Contack	(Waddington)
Hex	(Spear, Invicta)
Hexadoms	(Marx)
Maestro	(Pan)
Make-a-shape	(Taskmaster)
Multipuzzle	(Spear, Brightway, Invicta)
Pentominoes	(Taskmaster)
Polyominoes	(Invicta)
Super Puzzle	(Spear)
Tangle	(E.S.A.)
Tangled Angles	(E.S.A., Invicta, Taskmaster)
Tangrams	(E.S.A., Galt, Taskmaster)
Ten-up	(School Utilities)

Miscellaneous

Design Doodler	(Merit)

Section A2. Three Dimensions

Construction kits

Bilofix	(E.S.A.)
Meccano	(E.S.A., Galt)

Cube puzzles

Beat the Elf	(Waddington)
Construct-a-cube	(Galt)
Cube Fusion	(Waddington)
Kolor Kraze	(Waddington)
Soma Cube	(Invicta)

Making polyhedra

Geofix	(P. & T.)
Geometric Shapes	(Taskmaster)
Polyshapes	(Brightway, Invicta)

Networks and polyhedra

Construct-o-straws	(R.J.M., Arnold, Brightway, E.S.A., Galt)
Geostruct	(P. & T.)
Orbit	(R.J.M.)

Three-dimensional noughts and crosses

Four Score	(Spear)
Fours	(E.S.A.)
Space Lines	(Brightway, Invicta)
3D Noughts and Crosses	(Taskmaster)

Raw materials

Cardboard polygons	(Charles)
Cubes	(Arnold, Osmiroid, School Utilities, Taskmaster)
Geoboards	(Brightway, Cuisenaire, Galt, Invicta, Taskmaster)
Grids and Lattice Paper	(Altair, Copyprints)
Gummed Paper Polygons	(Charles)
Pegboards	(Brightway, E.S.A., Galt, Invicta)
Pegs	(Arnold, Brightway, E.S.A., Galt)
Squared Paper	(Arnold, E.S.A., P. & T.)
Squares	(Galt, School Utilities, Taskmaster)

Section A3. General

Self adhesive plastic sheet variously coloured

1. Velbex made by Bakelite Xylonite Ltd. (B.X.L.) obtainable from
   - (a) B.X.L. Southern Region Depot, (For Midlands & the South)
     Higham Station Avenue,
     London E 4.
   - (b) Trendella Ltd., (For the North West)
     Breightmet Beachworks,
     Bolton,
     Lancs.
   - (c) Sidney Beaumont, (For the North East)
     Croydon House,
     Croydon Street,
     Holbeck,
     Leeds.

2. Plastigraph

> Mathews Drew and Shelborne Ltd.,
> 78 High Holborn,
> London WC1.

3. Taskmaster Products (Leicester).

Slide rule demonstration aids

(Giant models, overhead projection models, etc.).

> (a) Aristo Technical Sales,
> Lupus Street,
> London SW1.
>
> (b) Blundell Harling,
> Weymouth,
> Dorset.
>
> (c) British Thornton Ltd.,
> Wythenshaw,
> Manchester.

Dienes apparatus

> Multibase Arithmetic Blocks (M.A.B.) ⎫
> Algebraic Experience Material   ⎬ E.S.A.
> Logical Blocks   ⎭

Cuisenaire rods and materials

> Cuisenaire Co. Ltd.,
> 40 Silver Street,
> Reading,
> Berks.

Gear wheels, pulleys, etc.

> Meccano, Bilofix and like construction kits.
> PROTO mechanical components (Miniature slotted angles,
>   gears, racks, chain wheels, pulleys, etc.) from:—
> Model & Prototype Systems Ltd.,
> 205 Kings Road,
> Kingston-upon-Thames,
> Surrey.

Reference

See also A.T.M. Pamphlet, *Materials for Mathematics* (A.T.M.) (5p).

Section A4. Manufacturers

Denys Fisher Toys Ltd.	Thorp Arch Trading Estate, Boston Spa, Yorks.
Louis Marx & Co. Ltd.	Swansea.
Merit Toys & Games	J. & L. Randall Ltd., Potters Bar, Herts.
Peter Pan Playthings Ltd.	4 Rodney Street, London N.1.
J. W. Spear & Sons Ltd.	Green Street, Enfield, Middx.
Tri-Ang	157 Edgware Road, London W.2.
J. Waddington Ltd.	Patrick Green, Woodlesford, Leeds, LS26 8HG.

Section A5. Suppliers

Altair Design	52 Borough High Street, London S.E.1.
E. J. Arnold & Son Ltd	Butterley Street, Leeds, LS10 1AX.
Brightway: Thos. Hope & Sankey Hudson Ltd.	123 Pollard Street, Manchester M1 2NH.
Charles & Son Ltd.	Woodbridge House, Clerkenwell Green, London E.C.1.
Copyprints Ltd.	87 Borough High Street, London S.E.1.
Cuisenaire Co. Ltd.	40 Silver St., Reading, Berks.
E.S.A. School Materials Div.	Pinnacles, Harlow, Essex.
J. Galt & Co. Ltd.	Brookfield Road, Cheadle, Cheshire.
Invicta Plastics Ltd.	Education Aids Div., Oadby, Leics.
Osmiroid Educational	Osmiroid Works, Gosport, Hants
Philip & Tacey Ltd.	North Way, Andover, Hants
R.J.M. Exports Ltd	Fairspear House, Leafield, Oxford OX8 5NT.
School Utilities Ltd.	5–9 Church Lane, Romford, Essex.
Taskmaster Ltd.	165/7 Clarendon Park Road, Leicester.

Section B1. 8 mm film loops

The following list is not comprehensive, but it will give an idea of what is available. All the loop films in the list are published by MacMillan. Those starred(*) are also available for use with Super 8 machines.

*SP/M/3   Uniform Linear Velocity.
*SP/M/4   Uniform Acceleration.
*SP/M/5   Force, Mass and Acceleration.

*SP/M/6	Motion at Right Angles.
*SP/M/7	Motion in a Circle. 1.
*SP/M/8	Motion in a Circle. 2.
*SP/H/3	The Reflection of Radiation.
	This film illustrates how reflectors with a parabolic cross-section can both transmit heat rays from a *focus* and also concentrate them at a focus.
US/6	Experimental Weightlessness.
US/7	Free Fall in Space.
	Includes film of passengers 'weightless' in aircraft.
SPC/1	Speed 1.
SPC/2	Speed 2.
SPC/3	Speed 3.
SPC/4	Instantaneous Speed 1.
SPC/5	Instantaneous Speed 2.
SPC/6	Instantaneous Speed 3.
SPC/7	Average Speed.
SPC/8	Measurement of High Speed.
SPC/9	Types of Motion.
SPC/10	Uniform Motion in a Straight Line.
SPC/11	Motion with Uniform Acceleration.
SPC/12	Free Fall.
R/4/1	Computers in the Sky.
R/4/2a	Computers in Industry. (1)
R/4/2b	Computers in Industry. (2)
R/4/2c	Computers in Industry. (3)
R/4/3	Computer Systems.
R/4/4a	Simple Computer Programmes. (1)
R/4/4b	Simple Computer Programmes. (2)
*R/9/1	NOT Gate.
*R/9/2	OR Gate.
*R/9/3	NOR Gate.
*R/9/4a	AND Gate.
*R/9/4b	NAND Gate.

Section B2. Film strips with tape-recorded commentary

(a) Introduction to the slide rule.
(b) How to use the slide rule.
from: British Thornton Ltd., Wythenshaw, Manchester.

Section B3. 16 mm sound film in colour (17 min).

Critical Path. On hire from: Central Film Library, Government Building, Bromyard Avenue, London, W1.

Reference

See also pamphlets published by A.T.M.
- (a) Film Pamphlet 1. (Film list 1968.)
- (b) Film Pamphlet 2. (Films and Film Making.)

7 Pupils at work—some topics for discussion by teachers

At one of the Project Conferences a request was made that the Project team supply some help and guidance to teachers who meet in Teachers' Centres, Colleges or like places. The response to the request was twofold: (a) the production of a *Teachers' Ideas and Resources Kit*, now out of print; and (b) a collection of coloured transparent slides accompanied by a taped commentary. These slides recorded pupils at work in a variety of group situations, and they are available on loan on application to: The Information Officer, The Schools Council, 160 Great Portland Street, London, W1N 6LL.

This set of slides was accompanied by a paper which is summarized here. It asks, but does not answer, many questions. Answers to these questions must be agreed before any serious development can be contemplated.

SOME TOPICS FOR DISCUSSION

1. Organization

(a) *Of the pupils*
On what basis are the groups (pairs, triples or more) formed? Should they be:

    (i)   broadly of similar ability and interests?
    (ii) of mixed ability?
    (iii) friendship groups?

How does one deal with incompatibility of temperament or of interests among group members?

Should the grouping operation be conducted by the teacher? or should it be left to the pupils themselves? or done cooperatively?

(b) *Of the working space*
Unless a suitably furnished and equipped space is available this is a

matter for adaptation and improvization. Can the existing furniture be easily and *quickly* rearranged into a desirable pattern? e.g.

(i) Flat table-top space for some groups.
(ii) Floor space for some groups.
(iii) Wall space for some groups.
(iv) Water, sink, or other services for some groups.

Whilst maintaining adequate circulation space both for teacher and for pupils.

Is there a possibility of some temporary use of a neighbouring corridor or some such similar space?

What mathematically educational purposes can wall space serve?

(c) *Of the working materials*

How are the materials required made rapidly and completely available?

(i) Workcards, worksheets, workbooks or whatever form of communication is employed.
(ii) The physical materials required for a given assignment.

Are the pupils themselves conversant with any filing system used? Discuss the pros and cons of storing the materials in the room or of trolley-feeding material from a central source.

2. Curricular

What proportion of the total time available for mathematics should be devoted to this kind of approach? The answers to this could well prove variable according to the objectives of the exercise; e.g. is it (i) merely an opportunity for pupils to extend their own, or the group's mathematical/scientific interests? or (ii) an exercise linked with a defined syllabus, in which case all pupils might be working on a common topic, it may be on areas, or on volumes, at one particular time? This pattern makes large demands on apparatus and equipment of one particular kind.

Is it sufficient for a teacher to sense the occasion for a full class lesson or discussion and then take appropriate action? Should the units of work be sequential or is an element of randomness necessary or desirable?

How is progression to be achieved in the work?

What forms of presentation of pupil work are acceptable?

Are double periods essential or merely desirable?

3. The role of the teacher

(a) As an organizer and collector of materials, references and the situation generally, i.e. in full preparation.

(b) In the work room as an ever present help to pupils; controlling enthusiasms or suggesting leads, or discussing a situation with a group or an individual pupil, as the need presents itself.
(c) As a recorder of pupils' work and progress. Implicit here is the matter of criteria to be used in evaluating progress.

4. Evaluation — what is to be looked for?

To what extent should the performance of a pupil be judged against:

(i) Externally imposed (adult?) standards?
(ii) The norm of the pupil's peer group.
(iii) The past performances of the pupil.

Some points which will probably be given different weightings by different teachers.

A pupil's:
(a) attitude to mathematics learning;
(b) ability to follow a set of instructions;
(c) ability to go beyond a set of instructions and therefore to conduct an open-ended investigation;
(d) ability to appreciate and to acquire the mathematical skills, techniques and content inherent in the given situation;
(e) manual dexterity in the use of apparatus;
(f) inventiveness and imagination in finding a solution, or different solutions to a problem; this is likely to be more concerned with physical situations rather than with mathematical abstractions;
(g) coherence, clarity and some degree of elegance in communicating the findings;
(h) ability to work unassisted by the teacher, or, conversely, how much help has the teacher had to give?

What other factors are relevant?

For each of these factors a pupil may well have a different rating in differing situations.

Are we satisfied with an overall impressionistic evaluation or should something more formal, e.g. a three-point or a five-point scale, be used? Should progress tests be used, and if so, how frequently should they be given? What forms should they take?

5. Solutions and suggestions

In the Project publication *Mathematical Experience* (Chatto & Windus, 1970) some teachers have presented their solutions or part-solutions to some of these questions. The book will provide further material for critical discussion.